FESTIVAL OF THE SPOKEN NERD

HELEN ARNEY+STEVE MOULD

The ELEMENT *in the* ROOM

Science-y Stuff Staring You in the Face

"The Element in the Room is in equal measure both profound and irreverent, brilliant, and daft. A veritable feast of the written nerd."
Jim Al-Khalili

"A delightful romp through science at every level, from how to make a cocktail using octoploid DNA to what the meaningless end of the Cosmos will be like."
Zach Weinersmith

CASSELL ILLUSTRATED

An Hachette UK Company
www.hachette.co.uk

First published in Great Britain in 2017 by Cassell Illustrated,
a division of Octopus Publishing Group Ltd, Carmelite House,
50 Victoria Embankment, London EC4Y 0DZ
www.octopusbooks.co.uk
www.octopusbooksusa.com

Distributed in the US by Hachette Book Group
1290 Avenue of the Americas
4th and 5th Floors, New York, NY 10104
Distributed in Canada by Canadian Manda Group
664 Annette Street, Toronto, Ontario, Canada M6S 2C8

ISBN 978-1-78840-013-8

Printed and bound in China.

10 9 8 7 6 5 4 3 2 1

Commissioning Editor: Romilly Morgan
Senior Editor: Pauline Bache
Copy Editor: Charlotte Cole
Art Director: Juliette Norsworthy
Typesetter: Ed Pickford
Designers: Ella McLean and Naomi Edmondson
Illustrators: Richard Wilkinson and Grace Helmer
Picture Research Manager: Giulia Hetherington
Production Controller: Meskerem Berhane

CONTENTS

FOREWORD BY MATT PARKER

The Other Nerd

I have been working with Helen and Steve for approximately ten years (rounded to the nearest whole decade) on Festival of the Spoken Nerd, and it has been an absolute delight to watch them bring all manner of scientific concepts to life, live on stage. My field of mathematics is easily conveyed: I set up the required equations and axioms and let the audience extrapolate their own way to understanding. But Helen and Steve need to guide the audience around every subtlety of experimental design and pitfall of misconception.

As a trio, we began by performing live shows but always wanted to expand out into something like a book so people could enjoy what we do in the comfort of their own home. We wanted a format for Festival of the Spoken Nerd where people could start and stop, enjoying math and science at their own pace. So we filmed a DVD.

When the possibility of a book expounding math and science in the Spoken Nerd style came along, we thought it could be a good companion to our filmed shows. But sadly I was busy writing *The Wrong Book*.

So it was decided that Helen and Steve should forge ahead with the science bits. They didn't need me, I would only hold them back. They could cover everything from self-experimentation to the end of the Universe, laboring over every word as they poured their infectious enthusiasm into this very book that you now hold. It is poetry about motion.

I, however, was able to generate my contribution far more efficiently. As well as this Foreword, I have also sent Helen and Steve 127 pages of densely populated math for the book. I really crammed it in. Much of the fun has been left out as an exercise for the reader. And what an exhausting exercise it is. My contribution is more of a mathematical marathon compared to Helen and Steve's scenic saunter through science.

So enjoy reading their bit. You will get to see what it has been like for me working with them over the previous 1×10^1 years (to one significant figure). This book really is just their brains smeared out on the page. I assume all of my work will be attached as an appendix at the end.[H1+S1]

Matt

H1 + S1 Oh yes, look out for our sequel, *The Equation in the Not Going To Happen*.

FESTIVAL OF THE SPOKEN NERD.
WHAT'S THAT ALL ABOUT?

It's hard to describe what we do for a living. On the surface, it's quite simple: we stand on stage talking—and sometimes singing —about science. And, crucially, people pay us to do it. Over the past seven years we've found a wonderful audience full of curious people who like to have their brains tickled. Some of those people are also under the influence of alcohol and jokes.

But this is a book!
Good point. We spend a lot of time searching for mind-tingling ideas, experiments, and stories, and now we've collected our favorite ones together on these pages. Some are interesting, some daft, and some a bit of both. All of them contain science-y stuff that's around you right now, but you might not have had a chance to think about it like this before. And you certainly won't be able to think about it in the same way again.

What kind of science-y stuff do you mean?
The "aha!" moments that inspire us. The science at our fingertips that we grabbed for as kids. The feeling when our brains fizz and pop with new ideas. The science that grew out from those humble roots and continues to weave through our lives.

Come on now! Be specific …
Ah, you mean, what's actually in this book? Well, we've started with the basics, looking at the piece of scientific equipment you know best … your own body, a mobile laboratory full of quirks and questions. We'll help you find answers first-hand in **Chapter 1**. Then we take a glance at what you put inside that body in **Chapter 2**, with some edible experiments and unusual facts about where some of your favorite food and drink really comes from. **Chapter 3** takes you to the neuro-nerve-center of it all, the big boss upstairs: your brain. We'll show you how to give it a gentle poke to find out if it's really as in control as you think it is. Taking a look at the world around us next, we've combed the periodic table for Cinderella elements, the kind that don't always get the attention they deserve. They're almost certainly in the room with you right now, and they're also in **Chapter 4**. Next, it's

time to add some collaborators! You shouldn't have to science alone if you don't want to. When you reach **Chapter 5**, invite some friends over and start a party with our step-by-step guides to home experimentation and science cocktail recipes.

The Universe takes center stage in **Chapter 6**, as we look at space from Earth's perspective, and at Earth from the vantage point of space. Get to know what's out there, and enjoy a bonus contribution from a very special guest "star." Finally, in **Chapter 7**, we'll hold your hand as we investigate the future technology that will be running, ruining, and redefining our lives. Because it's good to know at least a little of what to expect between now and the end of time. Sometimes we've dug deep, sometimes we've only scratched the surface. Either way, there's plenty here that you won't find in other science books.

How should I read *The Element in the Room*?

With your eyes. And also your brain. It's not just about reading though. There are experiments that you can try at home scattered throughout this book. You don't need to read everything in order, either, so feel free to jump around and find the things that interest you.

If there's something that you'd like to share while you're reading, find us online. We're on Facebook and Twitter: @moulds, @helenarney and @FOTSN. Take a look at our YouTube channels too. Corrections can be emailed to straight_in_the_trash@fotsn.com.

I want to do all the experiments!

Fantastic! But we're going to have to trust you on this. Basically, don't be an idiot. The most enjoyable experiments in this book are usually the most dangerous, so please stay safe. Take time to consider the risks before you start doing anything and remember that your safety is your responsibility. These experiments should not be carried out by children, or reckless adults. Several of them involve alcohol, or fire. Those experiments should not be conducted simultaneously.

WHO ARE HELEN AND STEVE?

 When you see this symbol, it means Helen has written this part of the book. You can recognize her from the glasses and lab coat. Although she has a degree in physics, Helen has never actually had to wear a lab coat to conduct any science experiments. So this is her only chance.

When you see this symbol, it means Steve has written this part of the book. Steve always wanted to be an astronaut. This book is the only way he's going to make that dream a reality—at least on paper.

Helen Arney
Helen is a science presenter and geek songstress with the "Voice of an Angle."[S1] You might have seen her explaining physics while riding a roller coaster for *Coast* on BBC2, electrifying the host of *BBC QI*, Sandi Toksvig, or hosting *Outrageous Acts of Science* on Discovery Channel. She can sing the entirety of Tom Lehrer's "The Elements" song—including all the new ones—and has filled several notebooks with rhymes for Uranus. None of them are printable here.

Steve Mould
Steve is the maker of mathematical curios, creator of illusions, and poster of quirky science experiments on YouTube. He is the author of *How to be a Scientist*, a book of fun experiments for kids, and—however ludicrous it sounds—has an actual scientific effect named after him (see page 141). You may also have seen him on TV in *Britain's Brightest*, *I Never Knew That About Britain*, and *Blue Peter*.[H1]

[S1] We mean that literally—she uses it to smash wine glasses in her spare time.
[H1] About ten years ago and he still won't stop going on about it.[S2]
[S2] By the way, these bits at the bottom are footnotes. You'll see a lot of these throughout the book.[H2]
[H2] So you can tell which of us has written the footnote, we have introduced a simple code. In front of each footnote number is a letter: either an S or an H.[S3]
[S3] I love codes! Which of those letters is mine?[H3]
[H3] Oh Steve ...

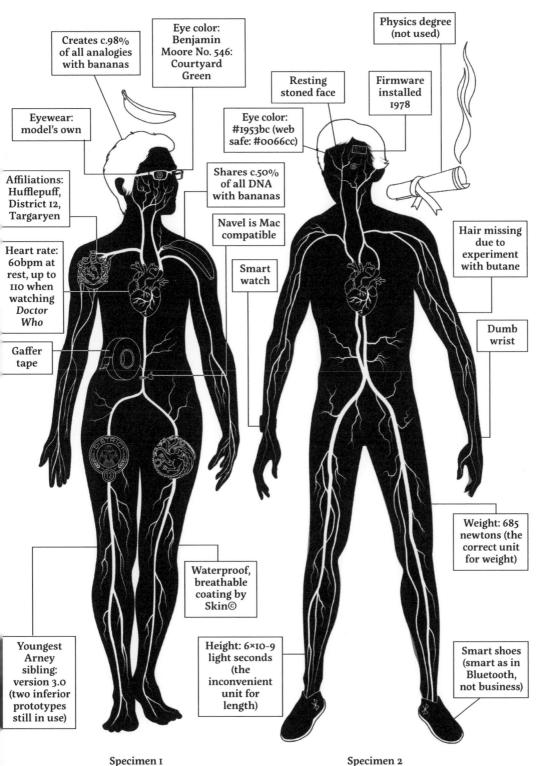

Creates c.98% of all analogies with bananas

Eye color: Benjamin Moore No. 546: Courtyard Green

Physics degree (not used)

Resting stoned face

Firmware installed 1978

Eyewear: model's own

Eye color: #1953bc (web safe: #0066cc)

Affiliations: Hufflepuff, District 12, Targaryen

Shares c.50% of all DNA with bananas

Hair missing due to experiment with butane

Navel is Mac compatible

Heart rate: 60bpm at rest, up to 110 when watching *Doctor Who*

Smart watch

Dumb wrist

Gaffer tape

Weight: 685 newtons (the correct unit for weight)

Waterproof, breathable coating by Skin©

Youngest Arney sibling: version 3.0 (two inferior prototypes still in use)

Height: 6×10-9 light seconds (the inconvenient unit for length)

Smart shoes (smart as in Bluetooth, not business)

Specimen 1

The "Helen Arney"

Specimen 2

The "Steve Mould"

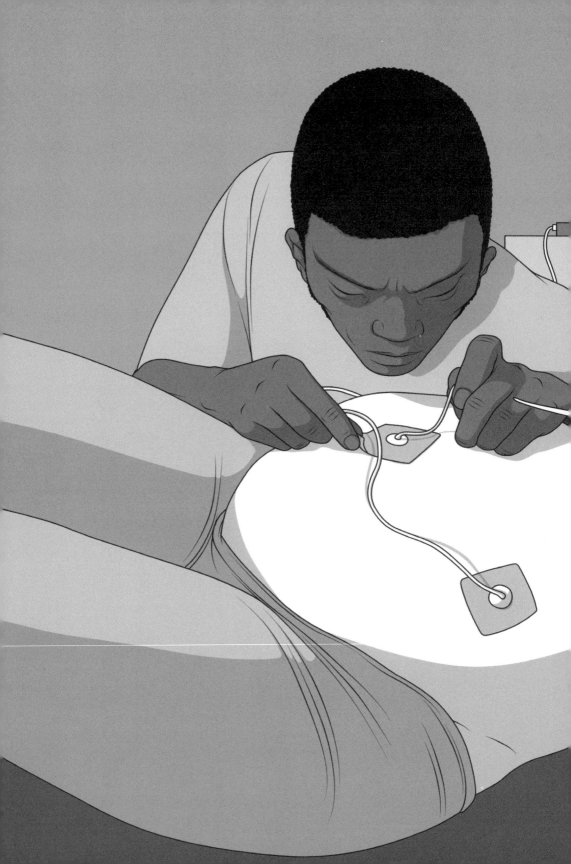

THE MIDWIFE WANTS TO KNOW HOW LONG THE CONTRACTIONS ARE.
AN ELECTROMYOGRAM SHOULD TELL US.
WOULDN'T A GALVANIC SKIN RESPONSE BE EASIER, DARLING?

BODY STUFF

1

Chapter

1

12/39

Human Bodies. We've all got one—unless you're a form of highly advanced artificial intelligence, in which case:

OIOOIOOO   OIOOOIOI   OIOOIIOO   OIOOIIOO   OIOOIIII   OOIOOOOO
OIOIOOIO   OIOOIIII   OIOOOOIO   OIOOIIII   OIOIOIOO   OIOIOOII   OOIOOOOI

For the rest of us, our bodies are the perfect place for a bit of home experimentation. Easy to access, infinitely fascinating, and—seeing as you already live in it rent-free—cheap. It's the do-it-yourself science laboratory you can play around in 24 hours a day.

In this chapter we explore a few of the human body's amazing quirks and qualities. We're all about getting to know some of the bacteria that live inside your gastrointestinal system, using physics to hear hidden sounds inside your head, and predicting the exact time of your child's birth, with ... a spreadsheet. And when you're done investigating your own body, get inspired by the animal kingdom and recruit your partner for some after-dark experimentation.

In we go ...

EXPERIMENTS YOU CAN DO ON YOURSELF

Find your v a l v e s

The blood in your arteries is pushed along by your heart, which means it's under quite a bit of pressure and there's no risk of it flowing the wrong way. Not so true for your veins, in which the pressure is much lower. To stop blood flowing backward in your veins, there are tiny little one-way valves all the way along, and finding them is a fun game you can play on your own or with a partner, or in a group—we don't judge.

The first thing to do is locate a nice big vein. If you've got veiny legs, this is your chance to shine. But you can also find good ones in your arms. For best results, allow your arm to swing down freely. Now place a finger firmly at one end of your veins, the end closest to your hand. This prevents blood flow in either direction. Now, press another finger into your arm, above but right next to the first finger. While still pressing, move your second finger up a couple of inches.

This evacuates the vein, so you should notice it go flat. But here's the cool part. Remove your second finger and blood will flow back into your evacuated vein, *but only up to the nearest valve.*

Congratulations, you've just found a valve! Interestingly, varicose veins are what you get when those little valves stop working.

Spin your foot and draw a 6

Place some cushions on a chair so that your feet would dangle if you sat down on it. If you have short legs you can skip this step.

Take a seat and start rotating your right foot clockwise, like so.

Now, with your right finger, try to draw a big 6 in the air.

The challenge is to complete the figure 6 without it messing up what your foot is doing.

It's almost impossible and you'll probably find your foot doing all sorts of mad movements.

This is the mating dance of the *LA Times* reader

Our brains have evolved to be good at controlling our bodies in rhythmic coordinated ways. It's what makes us good at walking and running. And it's very difficult to break out of, as this experiment shows.

Watch your eyes move

SPOILER ALERT

You can't.

Take a look in the mirror and glance back and forth between your left eye and your right eye. You'll be able to feel your eyeballs moving in your head as you do it. But you won't be able to *see* them move. That's because of a phenomenon called saccadic masking. When you glance from one thing to another, that fast movement of your eyes is called a saccade. If your brain were to process the visual information coming from your eyes during that time, all you would see is a blur. Saccadic masking is your brain blocking the visual processing while your eyes move. Your brain also does a great job of hiding the fact so you don't experience a little blackout every time you move your eyes.

If you do want to see your eye move, set your phone camera to selfie mode. There is a slight delay between the movements you make and those movements appearing on the screen, so you'll just be able to catch your eyes off guard. It's actually pretty freaky.

Floating finger sausage

Put your two index fingers together in front of your face (like ET would do if he were trying to heal his left index finger with his right). Make sure they're nice and straight and horizontal. Now focus your eyes on something behind your fingers. You should experience a "double vision" of your touching fingers. And it will appear as though you are now holding a weird sausage between the tips of your fingers. The sausage will have nails. Slowly move your fingers apart and you should see a floating sausage.

This tells us something really interesting about the way we process information from our eye and nothing interesting about sausages.

When you receive conflicting images from your eyes, like when you see double, your brain will suppress one of the images. So when one eye is telling you, "This is the end of your finger," and the other eye is telling you, "No, no. Your finger is still going," which eye wins and which gets suppressed?

In these scenarios, the high-contrast option wins, so it looks like your finger ends. In the space where the two fingers overlap, your brain concludes there is no conflict, only sausage.

Are you missing a tendon?

Place your forearm on a table with your palm facing upward. Now pinch your thumb and index finger together. And finally, bend your wrist so that your hand moves up.

There's an 85 percent chance you'll see a tendon pop up on your wrist.

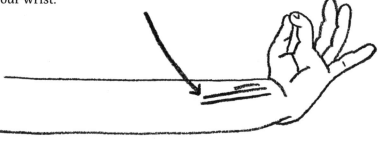

This tendon is connected to your palmaris longus muscle. A muscle that isn't in your body because *you* need it, but because your evolutionary ancestors needed it. It's a useful muscle to have around if you use your arms to scrabble around like monkeys, but humans don't do that anymore so the muscle is just a vestigial trait. Except, that is, for about 15 percent of the population. If that's you, congratulations, you have taken an extra evolutionary step, expect a phone call from the X-Men any day now.

Actually, this muscle and tendon are useful because they're useless. When a tendon or muscle is needed for reconstructive surgery, these ones are often used for spare parts because removing them has no effect on grip or mobility.

Lose all meaning

Try repeating a couple of words, like "Steve Mould," over and over again out loud. Eventually they start to sound like gibberish.

This is called semantic satiation because when you hear a word it causes a certain pattern of neurons in your brain to fire. These firing neurons are literally you understanding the meaning of the words: their semantics. If you've taken my suggestion, that would be something like "handsome clever man." As is so often the case in the brain, neurons that fire repeatedly become temporarily inhibited; they stop firing. This switches off your understanding of the words and they lose their meaning. If you repeat your own name instead of mine, it'll get very trippy indeed.

FROZEN fingers

Put your hands together like you're giving yourself a high five. Except don't follow through. Now fold down your middle fingers.

Try to separate your thumbs: no problem. You should also be able separate your index fingers and your pinkies without much difficulty (give yourself another high five, champ!).

What you won't be able to do is separate your ring fingers. Your fingers are pulled back by tendons in your hand and they all have separate ones except for the middle and ring finger so they can't be so easily moved independently of each other.

They say the fourth digit is used as the ring finger because it symbolizes the inseparable partnership of marriage. I don't know why they say that, though, because it's almost certainly not true. I mean, it really sounds made up.

Others say it's because there's a vein in your ring finger that runs straight to your heart. Unlike the veins in your other fingers that go via your butthole. That's not true either.

Feel an extra ball

1. Scrunch up a tiny bit of paper into a ball about the size of a pea.
2. Place it on the palm of your open hand.
3. Cross the index and middle fingers of your other hand.
4. Put your crossed fingers on top of the ball so it nestles between them.
5. Close your eyes and roll the ball between your fingers.

You should start to feel that there a two little paper balls in your hand. This is called the Aristotle illusion.

> "When the fingers are crossed, one object seems to be two; but yet we deny that it is two; for sight is more authoritative than touch. Yet, if touch stood alone, we should actually have pronounced the one object to be two." Aristotle

You are using two senses in this experiment. One of them is touch, clearly, but the second sense is less obvious. If you ask someone how many senses they have, most people will say five. But that's wrong. Scientists don't agree on how many senses humans have but they all agree that it's more than five. One of the senses not covered by the traditional list is proprioception, your sense of where your body parts are relative to each other.

Proprioception, however, is not *that* accurate and it doesn't cope well when you put your body parts in places they don't normally go (there is no joke here, just move along). It's also impaired by alcohol, which is why a police officer might ask you to tilt your head back, close your eyes, and touch your nose with your finger. You're much less likely to be able to do it while drunk.

So in our experiment, your brain doesn't account for the unusual position of your fingers and makes the assumption that they are uncrossed. In which case, the only way the normally outer edges of your fingers could be stimulated at the same time is if there were in fact two balls.

BE THE SCIENCE YOU WISH
TO SEE IN THE WORLD

I'm all for Steve encouraging you to donate yourself to science by finding valves and feeling balls. Self-experimentation is a long and noble scientific tradition, but sometimes it can go too far.

There's Albert Hofmann, the Swiss chemist, the first to extract LSD from poisonous mushrooms, and the first to experience an acid trip when riding a bicycle. It's a wonder the phrase "Don't hallucinate and perambulate" never caught on.

And there's Marie and Pierre Curie, who tested out their newly discovered radioactive elements in ways that included strapping them to their skin and waiting for burns to appear … some of Marie's cookbooks are still so radioactive they can't be handled without proper protection.

But my favorite cautionary tale is of Australian physician Barry Marshall and his pathologist colleague Robin Warren. In the early 1980s they disagreed with the general medical consensus that most stomach ulcers were caused by stress, bad diet, alcohol, smoking, and genetic factors. Instead Marshall and Warren were convinced that a particular bacterium, *Helicobacter pylori*, was the cause. And if they were right, the solution to many patients' ulcers could be a simple course of antibiotics, not the risky stomach surgery that was often on the cards.

Barry must have picked the short straw, because instead of setting up a test on random members of the public—and having to convince those well-known fun-skewerers of human trials: ethics committees—he just went ahead and swallowed a bunch of the little bugs.

Imagine the joy, as his hypothesis was proved right! Imagine the horror, as his stomach became infected, which led to gastritis, the first stage of the stomach ulcers! Imagine his poor wife and family, as the vomiting and halitosis became too much to bear!

Dr. Marshall lasted 14 days before taking antibiotics to kill the *H. pylori*, but it was another 20 years before he and Warren were awarded the 2005 Nobel Prize for Physiology or Medicine.

So, hang on, is self-experimenting really that bad if it wins you a Nobel Prize? I guess you can only have a go and find out … but please don't go as far as US army surgeon Jesse Lazear: in trying to prove that yellow fever was contagious, and that infected blood could be transferred via mosquito bites, he was bitten by one and died. The mosquito that caused his death might not even have been part of his experiment. It's thought that it could just have been a local specimen. But one that enjoyed both biting humans and dramatic irony.

Gastrointestinal elements

Something that my self-experimenting hero knows all too well is that no matter where you are, whatever you are doing, you are not alone.

You are home to trillions of bacteria, fungi, and archaea, living happily in and on your body. If you counted them all up, you'd find more of them than there are human cells in your body. Well, actually you wouldn't. Because even at the ambitious rate of one cell per second, it would take at least a million years to count them all.

So don't waste your time doing that, waste it instead on this little diversion. Here's 37 of the tiny life-forms inside your gastrointestinal system, set as a poetic tribute to all the life that you support without even thinking about it.

And, like the famous song "The Elements"—comedian and Harvard math professor Tom Lehrer's setting of the periodic table to music—you can sing it to the tune of Gilbert and Sullivan's "Modern Major-General Song," if you feel so inclined.

There's Peptococcus
Streptococcus
Fecalibacterium

Veillonella
Salmonella
And Fusobacterium

Plesiomonas
Pseudomonas
Also Eubacterium

Prevotella
Morganella
And Mycobacterium

There's Klebsiella
Eikenella
And Flavobacterium

Lactobacillus
And Bacillus
Propionibacterium

Citrobacter
Sarcina
Staphylococcus, Vibrio

Enterobacter
Bacteroides
And Butyrivibrio

Escherichia
(E. coli there ...)
And Corynebacterium

Helicobacter
Hemophilus
Bifidobacterium

Capnocytophaga
Don't forget Ruminococcus,

There's Methanobrevibacter
And Acidaminococcus

If you're feeling quite alone, just think of all the things inside of ya ...
Including Peptostreptococcus, Proteus, and Akkermansia

CAKE RACK GONG GONG GONG

Here's an experiment about a sound you can only hear inside your head.

To do it, you need to head into the kitchen and find a cake rack, a metal grill, or pretty much anything from your kitchen that's light and made of metal.

If you're using a cake rack, it's best to safely remove the cake first.

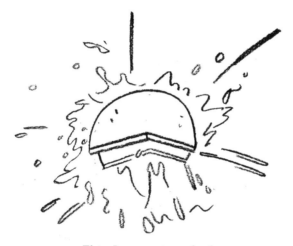

Fig. 1 Incorrect method.

Get two bits of string, about 12 inches long, and tie one to the top left corner of your cake rack, and one to the top right corner. You decide which corner is which. It's very important that you get it right.[H1]

Then tie loops in the other ends of the strings, big enough to pop one of your fingers in each, and suspend the cake rack in the air. Lean over to make sure it's not touching anything, as demonstrated by Steve overleaf:

[H1] It's not.

Now, take a metal spoon and rattle it across the bars of the cake rack. You may need to get a friend to help with this experiment, although if you're particularly keen on some solo science you can wedge your spoon under something so it sticks out over the edge of a table. It will need to be something heavy, like a cookbook or a bottle of wine. I often find that a heavy bottle of wine is a good substitute for a real-life friend.

Anyway, once you get your spoon rattling across the cake rack you'll hear it make a nice, pleasant, tinkly sound.

If you're doing this at home right now, it's likely that your whelm is very much under. That's fine. This isn't the actual experiment.

Because now comes the interesting bit ... Next you need to lean forward and maneuver your fingers up and into your earholes, with the cake rack still hanging down in front of you. Make sure it's still not touching anything. This should look like you're holding some sort of decorative rectangular stethoscope made entirely from kitchen implements. To visualize it, just imagine you're a you're a kid playing at being a physician, but only having stuff from the kitchen cabinet to make your imaginary stethoscope out of. Or take a look at Steve demonstrating it here again:

When you rattle your spoon against the cake rack now, it makes a completely different sound. It's one that only you can hear inside your head, so really, you should go to the kitchen right now and try it. If you can't, maybe because you're on a train, or in the bath, I can only describe it as sounding more like Big Ben than Betty Crocker. Instead of a tiny tinkle, you get a big booming sound.

Bonkers, huh?

There's a nice bit of science going on here, right at the junction of physics and physiology. It's clear that nothing about the cake rack has changed between your first and second attempts: the same vibration is making both the tinkling sound and the bonging sound. It's all about how that sound actually gets to your eardrums.

Most of the time we hear sounds coming to our ears through the air. But sound travels much more quickly through a liquid or —even better—a solid, and your head is the perfect material to make this happen.[S1]

What's going on $\mathcal{E}$ar then?

Air, it turns out, is a dreadful medium for sound waves to travel through. Here's a regular ear:

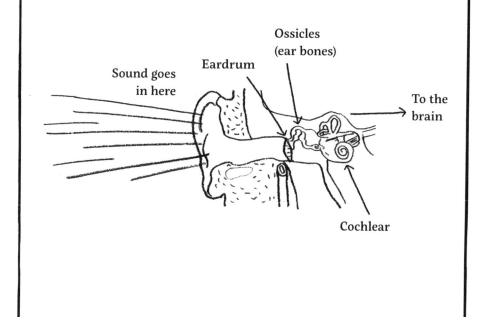

Sound goes in here

Eardrum

Ossicles
(ear bones)

To the brain

Cochlear

[S1] Are you trying to tell our readers that their heads are completely solid? That's pretty rude ...

Sound waves vibrate the air particles back and forth down the ear canal, transporting the sound through a series of nudges and wobbles. It's a surprise that any of it gets down there at all. As soon as the wave hits your eardrum, life gets easier. It bumps the ossicles, or ear bones, the three smallest bones in your body. Then it goes on to jiggle the fluid and tiny hairs of your spiraling cochlear. That turns sound into an electrical signal that zips up your auditory nerves and into your brain.

The weak link in all of this is the ear canal. When you connect the cake-rack gong to your head via your fingers, the vibrations don't just travel through the air. They go from the cake rack, up the string, through your fingers, and into the flesh and bones around your ear. You lose hardly any of the original vibrations, but instead connect directly to the internal apparatus of your ear so you hear a fulsome bong, instead of just a wee tinkle.

That's not the only sound you can hear inside your head.

Stick your fingers in your ears right now. Even without the string and cake rack, it sounds really … weird. You can hear the low pulsing sound of blood pumping through your body, the crunch of ear hairs being squished by your fingers, the sucking sound of your breath going in and out of your mouth and nose, the crack and pop of muscles and bones in your neck as you move your head around, the squelch of earwax as you adjust your fingers …

Take your hands away and it's gone. Thank goodness! All those bodily functions were freaking me out.

Well, it's not really "gone," unless something really dramatic happened to you when you pulled your fingers out of your ears. You just became more aware of it because you'd blocked any outside noises from entering your ear canal. And, at the same time, you'd "redirected" all your internal sounds back toward your eardrum by putting something solid across your earhole. It's called the occlusion effect.

This got me wondering: if you can hear those sounds from all over your body so clearly, do you really have to put your fingers *inside* your ears to hear the booming gong noise? The sound of a vibrating cake rack isn't coming from *inside* your body. Unless —again—something very dramatic happened when you pulled your fingers out of your ears.

The answer is no.

You could take my word for it, but seeing as you've gone to all the trouble of knotting string to the contents of your kitchen baking drawer, you might as well try it yourself.

Hang the cake rack from your fingers again, but this time wedge your fingers behind your ears, along your jawline, between your back teeth, at your temple, up your nose … Anywhere you stick your fingers transfers the vibrations through your head straight to your auditory system.[H1]

What you've just discovered is bone conduction. Listening to the real sounds of kitchen implements is obviously one of its key practical applications, but it does have others: hearing aids that link up to the bones of your skull, fancy headphones that sit just behind your ears instead of inside them, and waterproof speakers that let scuba divers communicate under the ocean. Too practical? Don't worry, in 2013 an advertising company tweaked the windows of a French train to play adverts to anyone who leaned against the glass, relying on bone conduction to get their message across.

Sleepy commuters beware! Not even the voices in your head are safe from the advancing tide of technology anymore. You could always take a stringy cake rack of your own and plug it into your ears mid-journey to cover up any rogue advertising messages. With the added bonus that no one will want to sit in the seat next to you.

[H1] You might want to use a pair of in-ear headphones or a fresh set of earplugs to stop sounds from the rest of the room overwhelming the more subtle bonging noise.

THE NERD'S GUIDE TO BEING A BIRTH PARTNER

I had a baby! And I learned a lot in the process. Having a baby generates a great deal of admin, most of it before the baby is even born, which is good because admin is hard with a newborn baby. One item on the agenda is putting together a Birth Plan, which outlines the various things you want to happen during the birth, such as who will be there, what position you want to be in, what drugs you prefer, and of course, what apps you'll use. As a birthing partner, you'll be involved in these decisions.

Apps

We used a variety of apps during the pregnancy, but I decided to limit myself to just one during the labor itself. Committing to a single app meant I was somewhat free to focus on other things, like being useful. The app I decided to use was a tracking app that recorded my wife's contraction times. I decided to go for a simple-looking one from the Android app store, but apps are available for the iPhone as well. If you've got a Windows phone, you can simply scratch the times into a rock with its sleek aluminum bezel.

I'M HELPING!

How does it work?

It's very simple actually, there's a big button on the screen that you press when a contraction starts and press again when it stops.

The app then tells you how long each contraction is and how far apart it was from the last one. For the uninitiated, these two nuggets of information are rather useful. It's like the signals coming from the alien mother ship in *Independence Day*. They get longer and closer together as time goes on and herald the end of the world. For accuracy, you might consider getting the mom-to-be to press the button. Though I found my wife, Lianne, was quite occupied with other things and didn't have time for my "stupid f**king app."

Why you should use an app

These numbers gave me a rough idea of how "far along" Lianne was, but I decided to check in with a health professional before driving to the hospital ...

A **water** birth

The big-ticket item on our Birth Plan was the pool, which is supposed to make labor a lot easier. And when we arrived at the hospital, it was there waiting for us. The only problem was, it hadn't been filled, which was an issue because Lianne was keen to jump straight in.

As her birthing partner, it was my job to persuade the midwives to fill it up, and this is when the app really came into its own. Let me explain …

The reason the pool had not yet been filled is that the midwives felt Lianne was not yet close to giving birth. So I brought out the app and showed them that, in fact, the length and closeness of her contractions suggested that she *was* close to giving birth.

But it turns out—and this will be a shock to you—midwives know more about delivering babies than I do and their assessment is more nuanced. In particular, they also look at how predictable the contractions are. Which is to say, do they come every one minute exactly or is it only one minute when you average out a whole load of wildly different contractions? Their own measurements suggested Lianne's contractions were still quite unpredictable.

I had a look around my app to see if it provided any sort of data like this. Perhaps it could tell me the standard deviation, for example? Sadly, no so such feature existed (is this a gap in the market?).[H1] But as I desperately scrolled through the menus, I spotted a glimmer of hope … an export button! I could download the data into Excel and analyze it myself. Thank goodness I remembered to bring my laptop.

When I told Lianne, I couldn't tell if she was happy or sad.

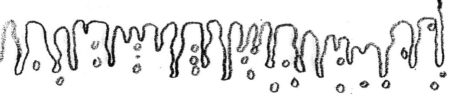

[H1] Definitely a gap in the market. Get on it, programming nerds!

I was able to quickly put a spreadsheet together. (I didn't know it but I'd been preparing for this moment my whole life), and I've reproduced below the graph I showed to the midwives.

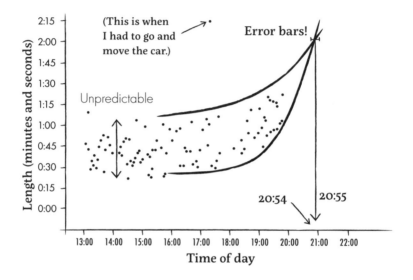

The little dots are the contractions and this plot shows what time they took place (x-axis) and how long they were (y-axis). The midwives were right! They were quite unpredictable. But you can also see that they are getting more predictable as time goes on: they are more bunched together to the right. I added the curved lines to illustrate this point. But more importantly, I extrapolated these lines into the future to show that at 20:55 they actually meet! This is the point at which my wife's contractions become perfectly predictable and, in my opinion, also the point at which my child would be born.

Despite this overwhelming evidence the midwives were still unconvinced. So I offered to show them my working out. At which point they filled the pool. Thanks contractions app!

In reality, my daughter was born at 20:54—one minute early. I was pretty annoyed as you can imagine, until I worked out the error bars in my prediction and found it was well within expectations. Unlike fatherhood.

So, was a nerd with a spreadsheet better at predicting the time of delivery than a team of midwives? I suspect I just got lucky. More research is needed. [S1]

Coincidence?

I like to think I'm a rational person. I like to think I have a good grasp of science and mathematics. I also think I have a solid understanding of probability; a good intuitive sense of what is likely and what is unlikely, informed by a rugged second-class degree in physics. So it was surprising to discover that Lianne had given birth to the best baby ever. This seems astronomically unlikely and yet is demonstrably true.

I did the only rational thing when faced with such an apparent conundrum. I began searching for other coincidences that would bolster my hypothesis that my daughter is somehow cosmically significant. It's what any scientist would do.[H1]

Our daughter is called Lyra after a Philip Pullman character, but Lyra is also the name of a constellation. You can look up on a clear night and see Lyra in the sky, and I had the rather brilliant idea of working out where Lyra the constellation was in the sky at the moment when Lyra the person was born.

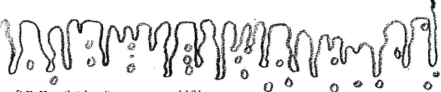

[S1] No Mom, that doesn't mean more grandchildren.
[H1] That's the exact opposite of what a scientist would do.

To pinpoint the location of a celestial object in the sky, you need the following things:

1. The ascension and declination of the object—these are its coordinates. They are independent of the orientation of the Earth so stay the same as the Earth moves.
2. Your longitude and latitude—these are *your* coordinates and they move with the Earth.
3. The exact date and time. This links the two coordinate systems together.

It's a complicated calculation, so I slaved away for hours … trying to find a website that would do it for me, which I eventually did.

And I discovered something remarkable. At the very minute Lyra the person was born, Lyra the constellation was directly above our heads.

Let that sink in.

At the precise moment my daughter was born, the very center point of the sky lay inside the constellation with which she shares her name.

If we were in a boat off the west coast of Portugal at the time. Which is even more amazing because Lianne speaks Portuguese!

If that isn't proof that Lyra is the best baby ever, I don't know what is.

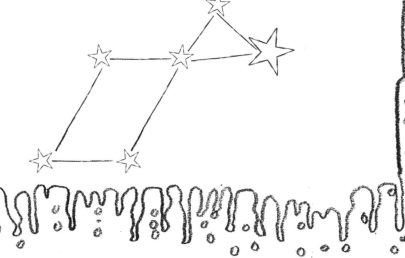

NERDS DO IT

Science is an inspiration all day long. And it's no less inspirational when it comes to that thing nerd couples do in the bedroom when the lights go out … Yes, Pierre-and-Marie-Curie role play!

Just me then? Fine!

I'm actually talking about what one scientist friend of mine calls "recreational procreation."

OK, I'm not being clear … I mean SEX! SEX! SEEEEEXX!!!

Sorry for shouting. I just find it difficult to talk about … you know … "it" … without using the safe, comfortable, and objective vocabulary of science. Which is where this section of the book comes in. I've come up with a guide to bedroom experimentation inspired by the animal kingdom and I hope that many readers will find this section of the book as inspirational as I do. Not so inspirational that you put down your copy and head upstairs with your partner. Wait until you've finished this chapter, OK?

So, when you're ready, switch your TV to the Discovery Channel and enjoy these pages for lovers … of the Natural Sciences curriculum.[H1]

Let's make love like bees

PERFECT FOR: A summer picnic.

During the act, the male's genitals explode. We've all been there. Am I right, ladies? Gents? Anyone and everyone?

It's actually a lot worse than I've described. The male honeybee, or drone, exists only for its

---

[H1] Remember to be safe, people, and always take appropriate precautions, whatever your area of research. For this section, a hazmat suit may be an unnecessary extra level of protection, but hey, whatever you're into …

chance to impregnate the queen bee. Which it does in midair, after competing with other drones in a game of high-speed aerial kiss-chase.

If that's already a tricky maneuver, wait for the next bit …

On piercing the queen's abdomen with its penis, the male's genitals explode and the poor drone falls to the ground in what is more of a "big death" than a *petite mort.*"

Although this is likely to ruin your picnic lunch, those kamikaze male bees have their reasons: it's a lot harder for the queen to mate again with the remnants of an endophallus sticking out of her. So the drone has just increased his chances of passing his genes on to the next generation of bees, while making the ultimate contribution to the art of cockblocking.

Meanwhile, thousands of worker bees carry on filling the hive with honey. Being female and unable to reproduce, none of them need to worry about exploding penises. Phew!

Let's make love like praying mantises

PERFECT FOR: A cheap date.

Instead of heading to a restaurant for dinner, try this budget version, where the female simply bites off the male's head and eats it as they get down to it.

This isn't simply kinky cannibalism, it's a brilliant technique that the female can use to get the most out of her encounter. The nerves responsible for the male's copulating motions are in his abdomen—not in his brain. Disconnecting the head removes any opportunity for the male to decide that his coital duties are done and it's time for a nap.

Yes, the male praying mantis doesn't actually need a brain to mate. If the female decides he's the one, she can just cut the small talk, lop off the head, and his body will just keep on going until he's, well, given all he's got to give.

A little myth-busting is needed here, though. Females often execute their male partners in captivity, but out there in the wild only around 30 percent of couplings end with an oral guillotine. The other pairings are more mutual, and both survive to WhatsApp their friends about it afterward.

So it's best to do this one alfresco, as your chances of "two becoming one" will be vastly reduced.

Let's make love like hedgehogs

PERFECT FOR: Less adventurous couples.

So how *do* hedgehogs make love? Carefully. And sneakily. The male hedgehog leaves a handy plug of semen in the female's vagina, making it far more difficult for any competitors to leave their mark.

Let's make love like salmon

PERFECT FOR: Cooling off on a hot day.

To make love like a male salmon, just do it in the bathtub and I'll drop by and pick it up later.

If you want a more true-to-nature version, simply substitute the "bathtub" for a freshwater spawning ground, and swap "doing it" for using your acute senses of smell and magnetoception to find the place where you were born. Yes, salmon use the Earth's magnetic field and unique scents from along the river to help find their natal spawning site.

If you can survive that ordeal without being eaten by a bear or dying from exhaustion, you're clearly a catch.

Let's make love like New Mexico whiptail lizards

PERFECT FOR: Putting the "sex" into "asexual reproduction."

New Mexico whiptail lizards are one of only a handful of species that have just females. Males have become obsolete. Instead of going to all the trouble of finding a mate of the opposite sex, the females reproduce through parthenogenesis, from the Greek for "virgin birth." An unfertilized egg cell divides and grows into an embryo containing only that female's DNA.

If that sounds like less effort than joining Tinder and waiting for someone worth swiping right for, you might be tempted to try it. Unfortunately, shrinking the number of genes in our collective gene pool by only making copies of ourselves would be a fantastically efficient way to end human life as we know it.

But these lizards have a secret weapon: they start off with twice as many chromosomes as their sexually reproducing cousins. By mixing and matching from a wider set of chromosomes, the next generation gets the same kind of genetic variety they would have got from two different parents.

Still, if the idea of attempting reproduction on your own appeals, you can always save this one up for vacations.

Let's make love like anglerfish

PERFECT FOR: Couples who want to take their relationship to the next level.

This one is kind of complicated but it's almost definitely worth it. You, the tiny male anglerfish, use your highly developed olfactory sense[H1] to detect my scent and swim for several days until you find me, the giant-size female smelling faintly of rotting meat. Already very sexy, yes?

And then you bite me.

That releases an enzyme on my skin that dissolves your scales, your flesh, and your … fins? Is that what they are? I can't tell in these low light levels in the deep ocean.

This leaves only a pair of genitals attached to the side of my body, for me to use when it's convenient. Of course, I'll be given many pairs of these "love gifts" from different partners over time, and I'll just collect them up as if they were shoes, or refrigerator magnets, or Pokémon[H2], until the time is right.

And if you find a really, really old female anglerfish, out there in the subzero temperatures of the darkest corner of the deepest ocean, you'll find that she is—quite literally—covered in balls.

Let's make love like pandas

Aaaaaaaaaaaaaand that's the end of this section.

Not really. The reputation of pandas as shy, awkward lovers isn't totally true—they seem to have no problem getting their fair share of good times in the wild. The problems mostly seem to crop up when they're popped into a concrete bunker and constantly watched by adoring zoo visitors.

I DON'T PERFORM IN PUBLIC …

Perhaps that's your thing? I'll not judge. Just don't make me pay an entry fee to watch.

[H1] Male anglerfish, in fact, have the largest nostril-to-head ratio of any animal. Impressive!
[H2] Gotta catch 'em all!

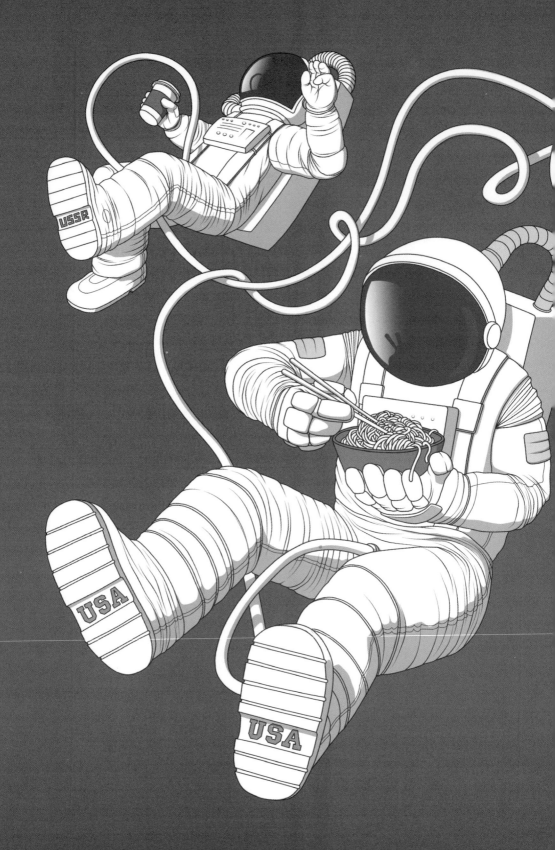

FOOD STUFF

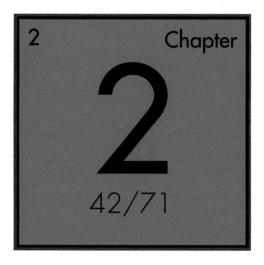

2   Chapter

2

42/71

Don't play with your food!

Shut up, I'm an adult now and I'll do what I like!

Playing with your food is a great way to bring out your inner scientist. You don't need to spend big bucks on fancy equipment, just open up your pantry and dive in. To help you get started we've prepared our favorite edible experiments, together with some bite-size chunks of fact-filled science, for you to serve up at any meal time.

Read on to discover the convoluted story behind your speedy morning beverage, via South America, an Alpine mountain range, and the International Space Station. On top of that we've got a new theory about the milk you pour into your daily cup, and an experiment you can do with the empty mug when you've finished drinking. And if you've ever wondered where minty-fresh flavors come from (spoiler alert: it's not actually mint) or how to win a nerd's heart on your first date, the answers are here too.

But before you get stuck in, make yourself a nice cup of instant coffee ...

GET TO KNOW YOUR CUP O' JOE

Coffee in an instant

If you're like me, you don't care how your coffee tastes in the morning, you're simply interested in the most efficient way to attach caffeine molecules to the adenosine receptors in your brain. And for people like us, there's instant coffee.[H1] It tastes like crap but you're not awake when it happens. The important thing is you can make it instantly and it's technically coffee.

Coffee Venn Diagram

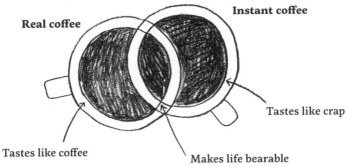

Real coffee

Instant coffee

Tastes like crap

Tastes like coffee

Makes life bearable

I should take a moment to address our American readers directly and explain what the hell instant coffee is. It's something you put in hot water that turns it into a caffeinated beverage. You wouldn't recognize it by its color, taste, or smell, but it is derived from coffee beans. There's a reason you may not have heard of it. After its invention, instant coffee only really took off in countries that didn't already have a strong coffee-drinking culture. For countries that didn't know any better, it was a big hit and people gulped it down with enthusiasm. It's coffee for amateurs, basically. A nice entry point for beginners. As Brits, we don't mind being labeled as such. We're traditionally tea drinkers, you see.

[H1] For an experiment that tests the effects of caffeine on your brain, see Chapter 5.

And although our coffee is bad, it's not as bad as your tea is bad.

Crap or Really Crap, it's your choice

There are actually two types of instant coffee. Crap and Really Crap. It's useful to be able to tell the difference when you're in the store. To tell the two apart, all you need is a sharp eye and a bit of history.

The story of instant coffee starts in Brazil. In the 1920s Brazil was producing about 80 percent of the world's coffee and it was a great business to be in because caffeine is addictive. It's the most widely consumed psychoactive drug in the world. Even during tough times, people need their fix. Things would need to get pretty bad before people stopped buying coffee and in 1929 they did. The Wall Street crash was the worst financial disaster in history and it left Brazil with a mountain of beans that were slowly going off.

Worried investors called upon Nestlé to invent a way to preserve the coffee. And if at all possible, retain some of the flavor too.

They devised a method to heat the coffee and dry it out. Perfecting the technique took roughly seven years, which in Brazilian terms is equivalent to about three or four forceful changes of government.

DRYING IT OLD S$_K$$_O$^{O}L

The process they developed was called spray-drying and it works the same way we dry most things: by heating them up. When you heat something up, you're really just making the atoms and molecules jiggle more. That's what heat is: molecular jiggle. And if you jiggle water molecules hard enough, they'll jump clean out of the thing you're heating and they'll evaporate.

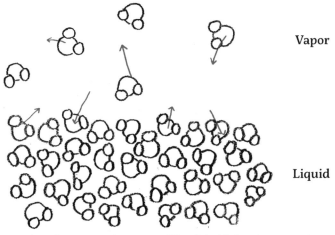

Vapor

Liquid

Water molecules can only escape like this if they're near the surface. The spray-drying process exploits this fact. First, make a batch of coffee, then spray the coffee liquid into a hot box. The spray creates a fine mist exposing more water to the surface.

Sadly, other molecules that give coffee its richness evaporate too. Goodbye flavor! So this is the Really Crap coffee I mentioned. Also called "instant coffee powder"—something of a misnomer because the powder is clumped together into tiny coffee poops.

If you'd rather your coffee tasted simply Crap, you'll have to wait until after the Second World War and the discovery of freeze-drying (oh, you did!), a technique used to preserve blood serum in the absence of reliable refrigerated transport. You can't dry serum with heat because people tend to be fussier about the chemical composition of their blood.

Let's freeze-dry all the things

Once discovered, freeze-drying was tested on loads of other things and it was found that freeze-dried coffee tasted WAY better than the spray-dried stuff.

Freeze-drying is a really weird way to dry something because instead of heating it up you cool it down. Here's what you do:

Make a really strong batch of coffee, so strong that it has a syruplike consistency. Pour it out into a thin layer, then cool it down to about -40°F (what's that in celsius? Also -40!). Once it's frozen, put your slab of coffee in an airtight box and suck out all the air. That's the key part. By sucking out the air you're lowering the pressure inside the box. And at low pressure water does something remarkable. It turns from a solid straight into a gas: a process called sublimation, and what you're left with is dry coffee chips. Skipping the water phase like this helps to keep those tasty aromatic molecules locked up in the coffee.

Step 1: Cool until the water in the coffee freezes.

Step 2: Lower the pressure.

Step 3: Warm up again and the water will turn from solid ice straight into a gas.

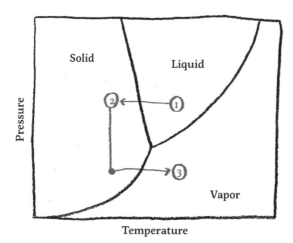

Freeze-drying comes with a great little side effect. It makes the instant coffee super-instant.[H1] To see why, let's look at drying fruit. Imagine trying to dry a raspberry by heating it up in a pan. Before it starts to dry it will go mushy. Heating it, then boiling the water inside, destroys its structural integrity. You'll end up with a hard, formless lump of fruit.

But freeze-dry a raspberry and it's a different story. The water molecules escape the fruit as a gas at a low temperature, leaving the structure of the raspberry intact. You often get fruit like this in cereal. It's light and fluffy, and full of little holes where the water once was. This makes the fruit really porous and ready to reabsorb water. The same goes for freeze-dried coffee, making it the most instant of all the coffees!

Freeze-drying happens in some unexpected places too, like in the freezer. It's just normal atmospheric pressure in there, so your frozen food dries very slowly. The result is called freezer burn but it's actually an example of freeze-drying. You may have read that you should only leave food in the freezer for a certain amount of time before eating it. That's not because it's dangerous, it's just because freeze-dried chicken doesn't taste that good.

And it's not even a modern invention. In 15th-century Peru, the Incas would freeze-dry their crops by lugging them up a mountain. The cold air and low pressure were perfect for a bit of sublimation. In fact, if you were to die in such a place, there's a good chance your body would be preserved for thousands of years. Like Ötzi here.

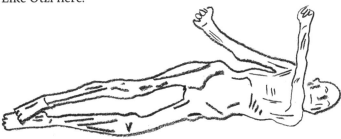

[H1] There's technically no difference between "instant" and "superinstant" as they are both "instant," but when Steve has had this much coffee, I'm not going to argue.

Ötzi is about 5,300 years old and still has skin and nails. Found in the Alps, he's referred to as a natural mummy and is the oldest of his kind to ever be found.

HA**PP**Y SHIT THANKSGIVING S^PACEM**AN**!

Freeze-drying is also great for reducing shipping costs. If you're sending a care package to a loved one, and you know they have water at their end, it makes sense to take the water out at your end before going to the post office. Especially if you're posting to the inhabitants of the International Space Station. Sending supplies into orbit costs somewhere between $2,000 and $10,000 per pound. On the Thanksgiving menu last year was freeze-dried green beans, mushrooms, and corn bread. The astronauts sent back their freeze-dried appreciation.

But where do astronauts get the water to put back in the food? Well ... it's recycled. So if you've ever bought "astronaut ice cream" from a science museum, you've not had the authentic experience unless you've rehydrated it with your own pee.

How to recognize the different types of instant coffee:

Freeze-dried—looks like light brown, little chips of wakefulness.

Spray-dried—looks like dark brown, knobbly clumps of powder.

MUG! THE MUSICAL

If that's put you off drinking any sort of instant heated beverage, don't worry. There's a nice bit of science that you can do with that empty mug and spoon in front of you. And, like a caffeine shot, it'll get your brain working a little bit faster (but without the midmorning cookie crash).

Hiding inside every mug is a musical instrument, made out of physics—and ceramic. This is technically materials science, but let's not be fussy.

One note samba

To find it, grab your spoon and tap the rim of the mug, directly opposite the handle.

A single note will ring out. This works best with fancy china, but you should be able to get any old average mug from the cabinet to join in.

If right now you're in a branch of Starbucks angrily mashing a paper takeout cup into a pulp with a cry of, "Why won't it sing?," I'd suggest ordering a nice, calming chamomile tea to go, and trying this again with a real mug when you get to the office. And maybe cutting down on your early morning intake of stimulants.

It takes two to tango

It takes two to tango

So, you've made one note, whatevs. We've all pinged a wine glass or—if you've made better life choices than me—clinked a champagne flute and heard it sing a single, sweet ringing tone. But surely two notes is the bare minimum you need to call something a "musical" instrument?

Here's where the science happens. Move your spoon one-eighth of the way around the rim of the mug, 45 degrees closer to the handle, and hit it again.

You'll hear a second, higher note this time. Have a play around and you'll find the perfect place to whack the rim with your spoon to get two distinct notes about a semitone apart. Ta-daa![S1]

How zhong has this been going on?

What you've just discovered here is the same physics employed by the zhong, a type of Chinese singing bell dating from around 3,000 years ago. Depending on where you strike it, the zhong plays two different notes. Indeed, a whole set of zhong in different sizes—a bianzhong—creates an entire orchestra of sound.

But it would be tricky to drink your breakfast beverage out of a cast bronze bell weighing somewhere between 4 and 440 pounds, so back to your mug.[S2]

[S1] Musical instrument? Two notes is borderline.
[S2] Having said that, coffee chains serve heart-stoppingly large cups of coffee as standard to create the illusion of value for money. So a Venti would probably fill a zhong or two.

Hit me *latte* one more time

The first time I saw this phenomenon was when I met up with my friend Colin for an early morning coffee. He showed it to me, and my mind was blown. Mostly because Colin had commandeered my empty mug for science before I'd managed to get any coffee into it, which—after a heavy night previously— had significantly lowered my threshold for mind explosions.

So how does it work? Well, the curves and nobbles of the ancient zhong should give you a clue about why your garden-variety mug produces the same effect. In a phrase that classical-music lovers like Colin will enjoy, it's a lot like Baroque music: all about the Handel.

There are two bits of physics going on here. When you tap opposite the handle, you're creating a standing wave around the opening of the mug—it's a wave that's "trapped" around the mug's circumference. It's this wave that vibrates the air around the mug and makes a sound that you can hear, and it's the frequency of this standing wave that dictates the pitch. On top of that, stuff you probably already know about the physics of musical instruments applies to mugs as well: bigger, heavier things make lower notes than smaller, lighter things when they vibrate.

But here's the thing: when you set up a standing wave in a circular object like a mug, not every part of it vibrates. The standing wave creates nodes—points that don't vibrate at all —and antinodes—points that vibrate a lot. And exactly where those nodes and antinodes appear will depend on where you hit the mug.

Tap your mug opposite the handle, and the handle gets in on the vibrating action because it's on one of those wildly vibrating antinodes. The extra mass of the handle makes the vibrating mug seem heavier overall, so it sings a lower note.

Tap it exactly one-eighth of a turn away, and the handle is on a node—it doesn't move at all. It's as if the handle has disappeared, the mug seems lighter and so it sings a higher note.

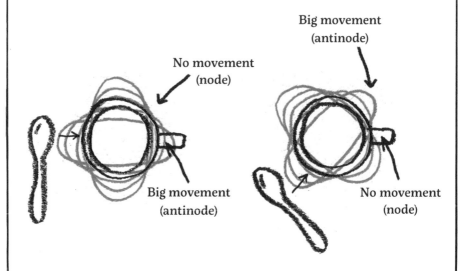

There are four places you can tap the mug to get the pattern of nodes and antinodes on the left, and four places you can get the pattern on the right. If you tap anywhere in between, the handle won't sit exactly on a node or an antinode, so you'll just get a dirty mush of various notes mixed together.

Now take a closer look at the zhong (on page 50). It does much the same thing: those little nobbles are in four sets of nine, with each set carefully positioned in perfect symmetry on the surface of the bell. They're not just decorative, they're like four sets of tiny handles that change the pitch of the bell depending on where you hit it.

And in case you're thinking that you can't play any actual tunes with just two notes …[S1] Well, yes you can. You can play my favorite tune of all time. Which happens to be the theme from *Jaws*.

MENTHOL

Two percent of people chew gum while reading, which is a surprising statistic given that I just made it up. But it also means there's a nonzero chance you're chewing gum right now. And if you are, it's most likely mint flavored. In which case, you've just experienced the great refreshing taste of … pine trees.

It's true! The *flavor* of mint comes mostly from a molecule called menthol, the world's most popular flavor molecule, with about 30,000 tons being produced every year.

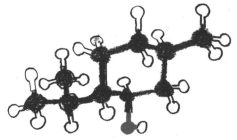

That's way in excess of what can economically be extracted from actual mint plants. In fact, you'd need to cover an area the size of Delaware in mint to supply the world's demand for menthol. Some people say that would be an improvement. Those people are from Maryland.

In fact, most menthol is synthesized in factories and the main ingredient is turpentine. You'll know turpentine as that bottle that's been under your sink for ten years. It's there for when you need to clean your paintbrushes. Which you never do. Which is why you keep having to buy new paintbrushes. Get your shit together. Or don't. In fact, if you turn that one-quart bottle of turpentine into menthol instead, you'll have enough mint flavor for about 200 tubes of toothpaste. And where does turpentine come from? Pine trees! So the first step toward minty fresh breath is milking a tree.[S1]

[S1] That might not be the right term, I'm not a farmer.

It used to be the case that synthetic menthol was inferior to the natural stuff, because it's hard not to simultaneously produce the mirror image of the desired molecule at the same time. The mirror image is decidedly unminty. In fact, it's usually described as musty. And it won't give you that cool sensation in your mouth, either, because it doesn't stimulate your cold receptors. These days, though, we're pretty good at producing only the desired menthol molecule without its embarrassing brother tagging along.

So if you're worried that your oral freshness isn't organic, it's OK, the synthetic method produces identical molecules. And molecules that are identical behave identically—unless you're into homeopathy, in which case, identical molecules do behave differently because of memory or magic, or something. Finally, synthetic menthol is cheaper to make, has a smaller carbon footprint and is better for the environment.

Phew. Everyone relax. Except, what about the rest of your chewing gum? What's that made of? This is going to sound bad, but it's made from crude oil. The thing that makes most gum chewy is drilled out of the ground. It used to be that gum was made of rubber extracted from rubber trees. But, just like mint, our insatiable desire for rubber meant a cheaper source was eventually found—from the petrochemical industry. Polyisobutylene is the final product and its production is less carbon-efficient than growing and processing rubber trees. But perhaps even worse is its stickiness. The cost of cleaning up chewing gum is so high that governments around the world are considering the introduction of a chewing-gum tax. It also means there's money to be made for anyone who can invent nonsticky chewing gum … Which means, before long, you may not be able to stick your chewing gum under the table at the library once you're done with it. You shouldn't be chewing gum in the library in the first place, to be honest. Sort your life out.

"I'D TAP THAT"

TIME FOR SOME CAN-NOODLING

If you're reading this book and wondering whether there's room in your own life for some home science, here's a confession: I actually ended up falling in love with and marrying my husband because of a kitchen experiment that went wrong.

It all started when we were both booked to play one of Robin Ince's science comedy nights at the Edinburgh Festival Fringe, having never met before. Both of us were doing stand-up about space, so it seems as if our meeting was written in the stars.[H1] Thanks pseudoscience!

But meeting is one thing; falling in love is another. I can identify the exact moment it happened and, this time, it was thanks to real science.

This is also where the noodles got involved.

It was during our first date, over a meal that started out as simply food, but turned into an experiment that we're still conducting today.

From Domestic Science to Domestic Bliss

My future husband, let's call him Rob (because that's his name), was busy cooking me his signature dish: chicken stir-fry. There he was at the stove, slaving away, instigating the Maillard reaction,[S1] denaturing proteins,[S2] and raising the temperature of some dihydrogen monoxide to over 373° Kelvin[S3] like any regular person, while I helped myself to a glass of fermented grape juice.[S4]

[H1] Our jokes were mostly written in the stars too. Well, about the stars. On an astronomical scale "in" and "about" are pretty much the same.
[S1] Let me translate: this means browning onions
[S2] Frying chicken.
[S3] Boiling water.
[S4] Although this sounds like a science-y way to describe wine, it was an actual bottle of grape juice that had been in the back of Rob's refrigerator for so long, it had unintentionally begun the fermentation process.

Rob had run out of his usual stir-fry accompanying carbohydrate: basmati rice, which he would delicately color and flavor with a spoonful of orange turmeric powder. Nom!

Here's where things started to go wrong, or—if you'd rather look at life as a scientist—this is where things started to become interesting.

So instead, Rob reached into the depths of his pantry to find a lonely package of ramen noodles. "No worries," he thought, "turmeric makes rice go a nice yellow color, let's just chuck that in with the noodles. Dinner is saved!"

And a few short minutes later our dinner was ready to serve.

At this point, it might be worth bringing up one of my favorite quotes from writer Isaac Asimov:

> `"The most exciting phrase to hear`
> `in science, the one that heralds new`
> `discoveries, is not 'Eureka!' but`
> `'That's funny …'"`

Or, in our case, not "Eureka" but:

> `"WHAT IN HELL'S MOUTH HAS HAPPENED`
> `TO THESE FREAKY NOODLES????!!!!"`

Because, dear reader, they had not turned a delightful earthy yellow color, they had turned blood red.

Aaaaaaaaaaargh!!!

Rob immediately tried three things—each of which had no effect whatsoever on the slightly alarming, bright red color of the noodles:

1. Rinsed the noodles in water.
2. Panicked.
3. Apologized profusely.

I immediately tried these things, all of which *did* work in neutralizing the slightly alarming, bright red color of the noodles:

1. Dipping the noodles in curry sauce—they turned back to yellow again.
2. Eating the noodles—thereby getting rid of the evidence.
3. Drinking the fermented grape juice—which for a moment made me forget about the evil death-noodle problem.

Now, some would suggest at this point that dinner was completely ruined.

Not us, no. For two nerds who had already discovered that they share a love of kitchen chemistry and Marie-and-Pierre-Curie role play, we looked deep into each other's eyes had the same thought:

"Dinner is now ... A Science Experiment!"

To find out what was going on here, we turned to the chemistry teacher we all wish we'd had at school: Google.

I spend a lot of time Googling science-y stuff. Looking back at my browser history, my last three Google searches were:

1. Experiments with cress.
2. Are Lucky Charms made by leprechauns?
3. Jim Al-Khalili.[S1]

[S1] Rob's last three Google searches are between him and his Internet service provider.

It turns out that turmeric has a chemical in it called curcumin[H1], and that curcumin is a natural pH indicator. It goes bright red when it's exposed to high pH alkalis[H2], and bright yellow when it's in a low pH acid.

Rice is slightly acidic, with a pH of around 6, so when you add turmeric, rice will go yellow.

But the noodles we were experimenting with[S1] were made to a traditonal Asian recipe that includes metal salts like potassium carbonate. That's what makes them taste so delicious, and also gives them a pH of between 9 and 11—well above the neutral midpoint of 7. So when you add turmeric, the curcumin reacts with the alkaline salts to turn them bright red.

If you then take one of those bright red noodles and dip it in curry sauce, it turns yellow. Clearly, curry sauce is acidic. That's not just a statement. That's an experimental result!

And so our date moved to the next level, as together we dipped and undipped our noodles into boiling turmeric water and curry sauce.[S2]

As we watched them turn from red to yellow and back to red again, we realized something. What we had basically made here, with our burgeoning love and our pantry ingredients was …

Homemade litmus paper!

Awesome.

So, of course, being "experimental lovers" of the less obvious kind, we couldn't help but try it out on as many household products as we could get our hands on—lime juice, baking powder, fermented grape juice, washing powder …

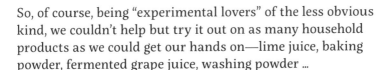

H1 Also known as diferuloylmethane to chemists, or E100 to its European food-additive friends.
H2 Technically we're talking about bases here, not just alkalis, but sometimes you have to contradict Meghan Trainor: it's not all about that base.
S1 The usual expression here is "cooking with."
S2 This is not an innuendo.

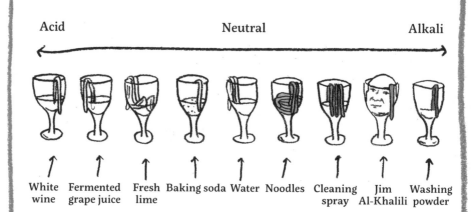

And display our results ... well, where else but on a table? The kitchen table.

As you can see here, acids are on the left, alkalis are on the right, and—just like Switzerland—the neutrals are in the middle.[H1]

So we completed our experiment. From hypothesis, through testing, to displaying results and now a conclusion, we had three important findings to state:

1. **We were still pretty hungry.**
2. There was no more fermented grape juice left.
3. Falling in love while you discover the pH of everyday household products is easy. Simply, **USE YOUR NOODLE**.

[H1] One of the most acidic of all the products tested was the fermented grape juice. It was only slightly less acidic than the $3.99 bottle of white wine from the corner store, which we later used to clean the bathroom.

PLAYING WITH YOUR PANTRY

Here are six of our favorite experiments for you to perform in your home laboratory. And by home laboratory, we mean that place in the house where you prepare food. Why, what do you call yours?

Put out a candle with an invisible blanket

1. Put about 3fl oz of vinegar in a pitcher.
2. Add a tablespoon of baking soda and put a plate over the top.
3. Light a few candles.
4. Once the bubbling in the pitcher has died down, remove the plate and pour the air from the pitcher over your candles.

This is the most romantic way a nerd can put out a candle.

A candle works by heating the wax until it forms a vapor. This vapor reacts with oxygen in the air in a process scientists call "burning." Burning produces heat which vaporizes more wax and so the process continues.

But if you can remove the oxygen from the air long enough for the wax to cool down, you can break the cycle and extinguish the flame. That's exactly what you're doing with the pitcher.

When you combine baking soda and vinegar, you're reacting sodium bicarbonate and acetic acid:

$$NaHCO_3 + CH_3COOH \rightarrow CH_3COONa + H_2O + CO_2$$

The products there are sodium acetate, water, and carbon dioxide (the bubbles). Carbon dioxide is heavier than air so will sit in your pitcher if undisturbed. Then, when you pour it out over a candle, you displace all the oxygen and the light goes out.

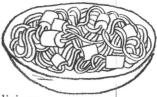

Impossibetti spaghetti

Next time you have a dinner guest, serve up this delicious sausage puzzle. You can either let them believe you slaved over a hot stove, individually sewing each piece of spaghetti, or challenge them to figure out how you did it.

The trick is simple, poke the spaghetti through the frankfurters *before* cooking it.

You might want to add some sauce too. I don't know, I'm not a chef. I hear tomato is good.[H1] And salt.

More impossible sewing

This time you're going to sew a guitar string through ice. What?!

The first step is to make some ice as normal. Well not quite as normal. Make a big old block of ice in a Tupperware container, for example, or an old takeout box.

Now place your block of ice on a table so part of it is overhanging the edge.

Get a metal guitar string or strip some wire down to the metal and attach a dumbbell (or some other heavy, even weight) to both ends.

Lay the wire over your ice so the dumbbell is dangling like so:

[H1] I recommend turmeric.

Best to do it outside in cold weather.

Over the course of a few hours the wire will pass through the ice. But, bizarrely, it won't cut the ice! So halfway through you'll be able to pick up the block of ice and show it to a friend, then present them with this puzzle: how did you make a block of ice with a wire running through it?

The wire is able to move through the ice because it is melting it. It's doing this in two ways.

The main way is through thermal conductivity. The parts of the wire that are outside the ice are at a higher temperature than inside, so heat is conducted through the wire into the ice. Once the wire has passed through, the melted ice (scientists call this water) above refreezes as it loses heat to the ice around it. Interestingly, as water freezes it actually releases heat to its surroundings. This is called the latent heat of freezing and because the wire is a good conductor, it will transfer some of this heat to the ice underneath it, speeding up the process.

There's a second, less pronounced phenomenon at play here called regelation. Water is unusual in that its melting point goes up when you apply pressure to it. It's a consequence of the fact that ice is less dense than water and it means that the wire pressing down on your frozen brick is actually raising its melting point. Regelation is also responsible for the pools of liquid water found at the bottom of glaciers. The water is literally squeezed from the mountain of ice.

A trio of breakfast experiments

I can't believe it's butter

You might have seen in the news about how inactivity is the new threat to our health. Well, here's an experiment that combines a delicious foodie treat with vigorous exercise: making your own butter from a jar of cream.

Find yourself a nice strong jar—preferably with a lid, or things are going to get messy—and pour it half full of whole whipping cream. The fattier the better! Check on the pot to make sure that for each 7fl oz of cream there's around 3oz of fat. Because that's where the science is hiding.

Cream is basically a mixture of fat particles and milk. But unlike salt or sugar, for example, those tiny, water-hating globules of fat don't actually dissolve in the watery milk. Those fat blobs don't much like sticking to each other, either. Instead, they just hang around, suspended in what's known as a colloid. The trick to making butter is to persuade all that fat to clump together. Usually, this is done by slowly churning cream, and it takes about half an hour. Boring! Some of the earliest recipes for butter suggest hanging an animal-skin pouch of cream from a pole, and swinging it back and forth until you get butter. Sounds more fun, but who has the time? Let's get cracking.

So, screw that lid on tight, and start shaking the jar, as hard and as fast as you can[H1]. The sloshing liquid sound will soon stop, but keep going[H2] until you hear a new, more watery sloshing

COLLECT ALL 300 EXPERIMENTS

[H1] Pro-tip: leave your cream out of the refrigerator for a while before you start. The closer it is to room temperature, the quicker it will churn.
[H2] If you stop here, you won't have any butter, but you will have started to break down the fatty molecules in the cream. Tiny air bubbles will be forming in between to create a nice big jar of whipped cream. Nommy!

sound start up again inside the jar. Don't stop now,[H1] you're nearly there! Keep on pumping!

Carry on shaking until you see a big blob of creamy butter sitting in a pool of watery buttermilk inside the jar.[H2]

Fish out that lump, rinse it in ice-cold water, squeeze all the extra liquid out, and treat yourself to a nice sit-down as you spread that hand-churned butter onto your bread.

In a game of calorie intake versus exercise, I'd call that a draw.

The element in your *breakfast* cereal

If you're a fan of breakfast cereals, here's an experiment that will whip your morning meal into a frenzy. Unfortunately it will also make it inedible, but hey, that's science for you. You'll need a really strong magnet for this experiment too—something much, much stronger than the one on your refrigerator. To get hold of a neodymium one, try the Internet. Don't forget to read the safety warnings when handling superstrong magnets. Mishandling can result in cuts, blood blisters, and—if you happen to be Iron Man or have a pacemaker fitted—death. And no one wants to deal with that at breakfast time.

Pour yourself a bowl of a breakfast cereal that's marked "fortified with vitamins and iron," and mix it with plenty of water instead of milk. Then chuck it into a blender to turn it into a delicious watery breakfast-cereal paste. Technically this could still be edible, but it's not a recommended serving suggestion.

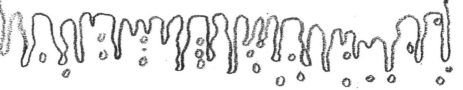

[H1] If you do stop now, it's still not quite butter, but you'll have broken down the fatty molecules even more. All the trapped air in that beautifully whipped cream will start to escape, leaving you with small yellow clumps of fat swimming around in a watery mixture. Not so nommy.
[H2] Hang on to this buttermilk as it makes good pancakes. Which are a great vehicle for your freshly hand-churned butter.

Pour your liquid cereal into a nonmetal bowl and let it settle for a few minutes ... then stick in your magnet and slowly stir it around for a few minutes more.[H1]

When you pull the magnet out, you should be able to see teeny-tiny, powderlike iron filings stuck to it.

Yup, iron filings. They were in your cereal all along! Best not to touch them, though, as loose iron filings can irritate skin and cause serious damage if you get them in your eye. And definitely don't eat them, now you've extracted them from their delicious-looking cereal mulch.

It turns out that something advertised as "fortified with iron" is more literal than you ever thought possible. Cereal makers add food-grade iron powder to some of their products, which in turn reacts with your stomach acid and lets you absorb some of the iron you need as it passes through your intestine. Your body contains about two small nails' worth of the stuff, so although it seems too weird for words, edible iron filings hidden in your morning meal are still better for you than anemia.

Now, where did I put that buttermilk? I think some breakfast pancakes will help me get over this ...

(Pancake) pan hovercrafts

As you whip up a batch of delicious buttermilk pancakes, get another experiment under your belt before you've even eaten by making some tiny hovercrafts out of water.

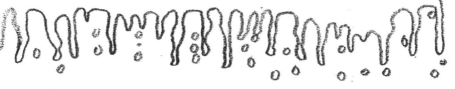

[H1] It's best to put your superstrong magnet into a clear plastic bag before poking it around in your mixture. That stops it from getting covered in iron powder, which then becomes almost impossible to remove. Trust me on this one.[H2]
[H2] It also makes the iron easier to see, as you can pull the plastic away from the magnet to take a closer look at the powder as it squelches around on the outside of the bag.[H3]
[H3] On an unrelated topic, if anyone needs a neodymium magnet covered in tiny pieces of iron, please do get in touch.

Heat your pan to a moderately high temperature, one that you'd be happy cooking pancakes in. Drip some drops of water into the pan and, provided that the temperature is about 212°F, the water will spread out and boil away. If it's just above 212°F, the water will evaporate almost immediately. Just like you'd expect. So far, so unscience-y.

Now crank up the heat to a stir-fry temperature, that's around 375°F. But how can you tell that your pan is around 375°F without using some kind of industrial device to measure the temperature? Simple. If this experiment works, your pan is around that temperature. And if your pan is around that temperature, this experiment works. I'm calling it "metaphysics."[H1]

So, make sure the pan is dry, the water you're using is clear, and this time when you drip water into the pan, the drops will dance around for a few seconds before they disappear.

Congratulations! You have instigated the Leidenfrost effect, named after the 18th-century German doctor who first described it. You've basically made tiny water-droplet hovercrafts in your pancake pan. Because the surface of the pan is much, much hotter than the boiling point of water, the droplet doesn't boil away immediately but instead forms an incredibly thin layer of water vapor between the blob of water and the pan. This cushion of steam lets the unboiled water "hover" above it, and also works as an insulating layer to stop the droplet boiling off right away.

It's best to let your pan get back to a reasonable temperature before you pour in the pancake batter—unless you like your hover-pancakes burned to a crisp.

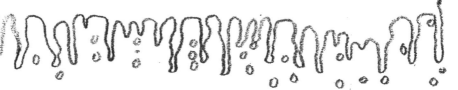

H1 This is how professional chefs work it out, anyway.

I FOR ONE WELCOME

OUR BOVINE OVERLORDS

In the battle for the survival of the fittest, there are lots of ways to be fit. Obvious ways, like being fast and having strength. But there are less obvious way too, like knowing how to motivate a workforce.

Take aphids, for example. Aphids hire bodyguards to keep them safe, so they need to know a thing or two about job satisfaction. Aphids are very good at extracting sugary sap from plants, but they're quite slow and vulnerable. So many species enlist the help of ants. In exchange for the aphids' excess sugar, the ants protect them from predators and parasites. Ants are big and tough, but not good at the whole sugar-extraction thing, so it's a good deal for the ants too.

But wait … are the aphids employing the ants like bodyguards or are the ants farming the aphids like crops? Well, neither and both. It depends how you look at it. Scientists have observed ants transporting aphids to new plants when the sap runs dry, and a single ant colony can maintain a large field of aphids. Some aphids have even lost the ability to poop sugar without

the help of ants stroking their abdomens. So you could argue that the ants have been selectively breeding these aphids by only protecting the best sugar producers. Or you could argue that aphids have been selectively breeding ants because their sugar offering only goes to the best protectors and farmers.

This two-sided story is told again and again throughout the natural world. My favorite example involves humans and dairy cows. Surely we are the ones in charge in this relationship. We selectively breed cows to be super milk producers. But it's a good deal for the cows too (or at least it's a good deal for the cows' genes). The world's dairy-cow population is huge, in part, because they produce something we want. What an ingenious way to get your genes into the next generation—persuade another animal to farm you!

But to begin with, humans weren't the ideal cultivator for cows. You see, humans can produce an enzyme called lactase that helps them digest lactose, the primary sugar in milk. But back in the day, we only produced lactase during infancy while we were breastfeeding and lost the ability after weaning, so cow's milk wasn't such an attractive product. Fortunately, some humans with a particular gene mutation kept producing lactase into adulthood and cows made sure these humans did well by supplying them with plenty of energy in the form of milk. These mutant humans were therefore slightly more likely to pass their genes onto the next generation and the mutation spread. Lactose tolerance (or more correctly lactase persistence) is now found in the majority of Europeans and 100 percent of people of Irish descent. That's right, cows have been selectively breeding humans all along. So next time you grab a bottle of milk from the refrigerator, just remember who's in charge.

BRAIN STUFF

3   Chapter

3

74/101

If we've learned anything from neuroscience and psychology, it's that people don't think the way you'd think they think. And their senses don't sense in a sensible way.

In this chapter we'll be busting some brain myths, showing how to twist your senses to your advantage, and differentiating your imposter syndrome from your Dunning-Kruger effect. All in all, this will be a journey of self-discovery, which should be easy once you've discovered the personality test to best reveal your personality.

But first, let's dig into the deep workings of your brain via your eyeballs with some choice optical illusions ...

EYE TRICKERY

I collect optical illusions like stamps, only they're in a folder on the desktop on my computer as opposed to a folder on my desktop on my actual desk. I love the classics, of course, like the two tables that are actually the same size, or the duck that's also a rabbit, or the vase that's actually twins getting all up in each other's grills.

The tops of these tables are the exact same shape and size.

Duck or rabbit?   Vase or twins?

But what I want to show you in this book are some of my favorite rare specimens, starting with …

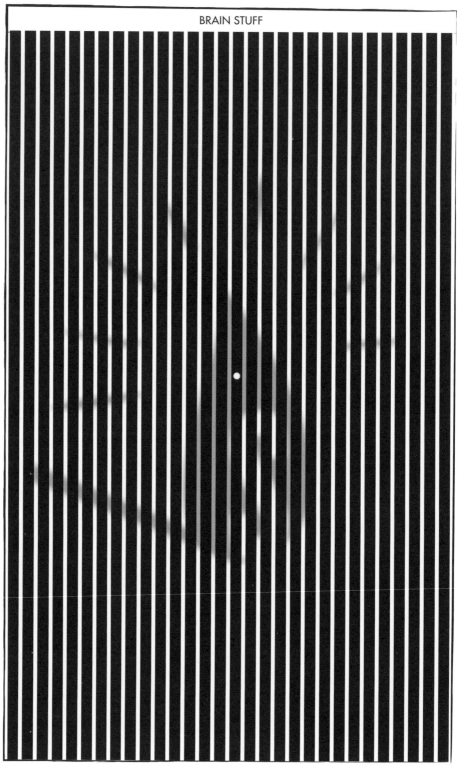

The ShakY head illusion

If you focus on the dot in the middle of the image, all you will see is black bars. But if you shake your head from side to side, you should see the hidden picture between the bars. Don't do this in public, people might think you don't like our book.

Two types of neuron in the visual system are responsible for the effect. Parvocellular cells (P-cells) and magnocellular cells (M-cells). P-cells are good at distinguishing fine detail, like the sharp edges of the black bars (they have high spatial resolution), but they perform poorly when things are moving fast (they have low temporal resolution) or where the contrast is low. Whereas M-cells aren't so good at fine detail (low spatial resolution) but can handle a bit of movement (high temporal resolution) and low contrast.

So when you fix your eyes on the dot, your P-cells are in charge and all you see is black bars. But when you shake your head, the M-cells are in charge and you can pick out the faint soft-edged image underneath.

Take a long look at the image that opens this chapter (on pages 72–3). This is not a spiral.

Neurons that interpret orientation in your visual field incorrectly combine the disconnected tilted segments into a continuous line. This happens early in your visual system before object recognition can kick in to pick out the circles.

The watercolor illusion

You may have spotted this book is black and white with one other color. Here are a couple of illusions that lend themselves nicely to this aesthetic. The first is this watercolor illusion:

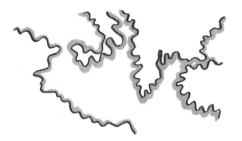

The "land" in this picture should appear to be a pale pink color but it's actually white.

The watercolor illusion also solidifies our perception of an object in front of a blank background. Without the pink inner border, the squiggly line seems more abstract and less like the outline of an object, as shown below.

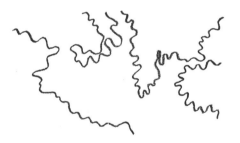

The watercolor illusion is not well understood but some interesting research has been done. For example, scientists have found the illusion still works when the outer dark line is sent to one eye and the inner light line is sent to the other eye. This suggests the illusory color is generated somewhere after stereo depth perception in the visual system.

The Munker-White illusion

The red you see throughout this book is called Pantone 185 u—a red that is slightly on the light side and nudged a little toward magenta.[S1] The Pantone system is designed to make it easier to standardize color reproduction in print by specifying an exact mix of 14 different pigments. It's a noble aim, but in the real world our perception of color is so heavily influenced by context that getting your ink *just so* isn't desperately important. This illusion illustrates that fact nicely.

The red stripes on the left are the exact same color as the red stripes on the right. Moving the book away from you will enhance the effect. This is called the Munker-White illusion, and its explanation is contested.

Interestingly, brightness illusions can usually be explained by lateral inhibition—the tendency of excited neurons to dampen surrounding activity. So a color with a bright surrounding would seem less bright than it really is. But we see the opposite in the Munker-White illusion! One possible explanation is belongingness. We perceive the red stripes on the left as belonging to the black stripes and so in contrast they are brighter. Whereas the red stripes on the right belong to The White Stripes, so appear darker by contrast and more interested in garage rock revival.

[S1] I know you want the hex value: #F15060.[S2]
[S2] OK, here's the CMYK value too: 0 81 54 0.

PERSONALITY TEST TEST
WHICH PERSONALITY TEST SUITS YOUR PERSONALITY THE MOST? TAKE THIS SIMPLE QUIZ TO FIND OUT!

Personality tests are popular with employers, recruiters, and click-happy Internet users who want to discover their spirit animal in a hurry. But they haven't always been based in actual, you know, evidence and stuff. In spite of that, research conducted by the Society for Human Resource Management suggests that 22 percent of US companies use them in the recruitment process, and the industry itself is valued at upward of $500 million a year.

In an attempt to divert some of that sweet, sweet brain-quizzing cash back into real science, here's the personality test to beat all personality tests.

Question

Which of the following methods of discovering someone's personality type do you prefer?

A. Roll up your sleeves, oil up your hands, and feel for bumps on their head.

B. Ask a series of carefully worded questions, then analyze the heck out of their answers.

C. Spill some ink and ask them what it looks like.

D. Look up the moment and location of their birth on an ancient chart.

E. Compare them to the main characters of a series of fiction books set in a boarding school for wizards.

F. None of the above.

YOU ANSWERED A

Why not try ... **phrenology?**

Yes, that old chestnut: the idea that the size, shape, and weight of your skull can indicate the good and bad qualities of the mind inside. The theory was developed in 1796 by Franz Joseph Gall, who believed that the brain was made up of 27 different parts, each with specific functions that influenced a person's personality.

Even nowadays the concept appears to have a smidgen of truth about it, because we know today that the brain does indeed have different regions with specific functions. Gall drew a map and everything, which made his idea look properly science-y, and also quite beautiful. You might have seen it printed on expensive mugs in museum gift stores.

Gall even predicted that other animals would have similar functional regions, or "organs," in their own gray matter— an idea that rings true as we learn about how brains work in other animals.

But in 1825, when French physician Marie Jean Pierre Flourens actually tested the location of Gall's "organs" on pigeons,[H1] it appeared that Gall had just made it all up. Throw your gift store china aside! Phrenology is total poppycock.

Quite aside from the half-science, half-bonkersness of it all, there was a dark side to all this head stroking. Phrenology became a standard cover for anyone who wanted to "scientifically" justify their own racist ideologies. For example, different skull shapes from across the globe were cataloged by some into their supposed stages of "evolution." Yeuch.

So why was it so popular? You can put that down to the 19th-century general public's enthusiasm for science entertainment. Phrenology lectures made this weird hobby of feeling bumps

[H1] Testing the internal workings of the brain on human subjects being somewhat challenging, until the advent of functional magnetic resonance imaging, or fMRI.

to reveal the mysteries of the human brain seem so much like science, but so much simpler to understand. Dammit science! It's still all your fault after all!

There's only one good reason to get into phrenology nowadays, and that's to receive a sensual head massage from a total stranger in a pseudomedicalized setting. If that sounds like fun to you, it probably says more about your personality than any scientifically rigorous test.

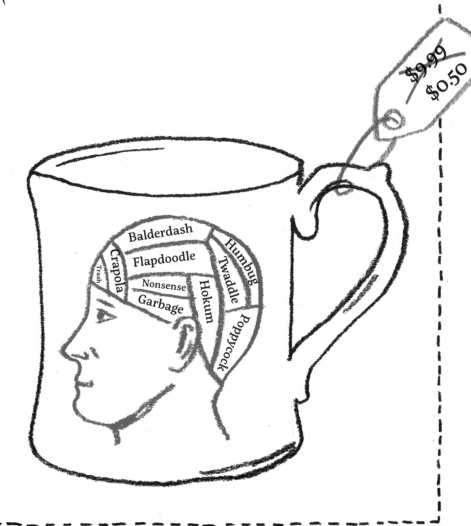

YOU ANSWERED B

Why not try ... **the self-report inventory?**

This is more familiar territory. You might have done one of these online, for a job interview, at an employee team-bonding weekend ... perhaps the MMPI,[H1] the MBTI,[H2] the TIPI,[H3] the FIPI,[H4] the IPIP,[H5] the DiSC assessment,[H6] and plenty of other FLA.[H7]

One of the earliest tests, the Woodworth Personal Data Sheet, was intended to screen newly drafted US Army recruits during the First World War. Asking for simple yes or no answers to a series of questions, it was designed to assess a subject's risk of shell shock by identifying recruits who were already suffering from some kind of psychiatric disturbance.

Also called the slightly less inviting Woodworth Psychoneurotic Inventory, it included questions like:

○ Does the sight of blood make you sick or dizzy?
○ Are you happy most of the time?
○ Do you sometimes wish you had never been born?

Cheery stuff. Unfortunately for its creators (but not for the rest of the world), the war ended before it could be used. No matter! The field of psychological research bravely took up the baton and the WPDS became the daddy of all personality tests.

The general idea behind WPDS still hasn't changed today, and now those little click-boxes are everywhere.

[H1] **Minnesota Multiphasic Personality Inventory.**
[H2] **Myers–Briggs Type Indicator.**
[H3] **Ten-Item Personality Inventory.**
[H4] **Five-Item Personality Inventory—"half the personality, twice the fun!"**
[H5] **International Personality Item Pool.**
[H6] **Dominance, Influence, Steadiness, and Conscientiousness.**
[H7] **Four Letter Acronyms.**

Some versions are more scientific than others in their approach and analysis, but that doesn't stop lawyers using them for criminal investigation, colleagues using them on each other to understand how to work together, and salespeople using them on their clients to understand how to sell stuff to them. Even singletons use them to increase their dating efficiency by narrowing in on only the most compatible potential partners.[H1]

But there are a few problems with self-reporting. They rely entirely on the questionnaire-taker's ability to look at themselves objectively. Most of us find that difficult. It's all too easy to slip into answering questions about our preferences from the perspective of our "ideal" selves. And on any given day we might answer some questions one way, other questions another. It's not unusual to find your personality type has "changed" if you take a second test several weeks after your first.

A bigger problem for those trying to make important decisions based on the tests is honesty. If a job candidate is given a self-reporting test and wants to "game the system" they potentially could, by answering questions in a way that gives them one of the personality types that suits the recruiter's criteria. It's a risky strategy, because the questions and methodology of most popular personality tests are closely guarded corporate secrets. If you don't game it right, you might still end up with the "wrong" type for the job and your erratic answers could raise a red flag that wouldn't otherwise exist on your application form.

But there are now so many people using self-report inventories for work or for pleasure—around 2.5 million people take the MBTI test every year—that you might as well give in and take every single one you can get your hands on. To help your friends and colleagues modify their language and approach when interacting with you, tattoo the results on your forehead for easy reference.[H2]

[H1] If your ideal partner absolutely must have the quality "enjoys filling in surveys" I can guarantee it really does work.
[H2] But my Myers-Briggs type is ENTP. So I would say that.

The dribble of compassion

The splodge of destiny

Your mom

Call of Cthulhu

You are feeling very sleepy...

YOU ANSWERED C

Why not try ... the *Rorschach* test?

We've probably all seen them: the ten iconic ink splotches of the Rorschach test, created by the Swiss psychologist who gave them his name. Hermann Rorschach never intended his blots to be used to assess personalities, but as a way to diagnose schizophrenia, and first published them under the unassuming title *Psychodiagnostik* in 1921. He never lived to see his namesake travel across the globe as he died the following year.

And yet here we are! In a world where forensic psychologists in criminal investigations, child psychologists in schools, and psychology professors in their lunchbreaks spend a fair portion of their time holding out the same ten cards with the same ten splodgy patterns and asking people what they see. Crabs, bats, dogs, and rude parts of the human anatomy[H1] mostly. Since its heyday in the 1960s, the Rorschach has fallen from favor in the US, UK, and most of the Western world, but is still strangely popular in Japan.

Whatever its validity, if you—our nerdy reader—try the Rorschach test, it may be wrecked before you even take it. Popular culture has already exposed many curious minds to those ten classic inkblot images and the most common interpretations, so it's possible that the entire test has been rendered invalid. That's why we have produced for you a "special" version of our own. We're just trying to keep things scientific. By being unscientific. Analyze that, Rorschach!

[H1] Because everything looks like a rude part of the human anatomy if you look at it for long enough.

YOU ANSWERED D

Why not try ... **natal astrology?**

Or don't. A double-blind test in 1985 by American physicist Shawn Carlson has settled the question of whether a birth-chart interpretation or horoscope reading can genuinely predict a personality by drawing conclusions from the position of the sun, moon, and planets and their relationship to each other. It can't. There haven't been all that many scientific studies (I mean, how many do you need?), but the ones that have been done show astrological predictions are no more accurate than chance.

Although it's widely regarded as Taurus-poop, an interesting phenomenon did crop up in 1955. French astrology researcher Michel Gauquelin found a correlation between successful athletes and the particular position of Mars on their natal charts; he called it—in a name lacking the creativity of most astrology enthusiasts—simply the Mars effect. Further analysis resulted in a split into two camps: some contradicted the findings, some backed them up, and the only definite conclusion is that the results are "inconclusive."

One possible explanation for the effect is interesting to consider. Rather than Gauquelin massaging his data to include certain sportspeople and exclude others, as he was accused of at the time,[H1] the cause may be something altogether more human. The original study looked at children born at a time when astrology was more popular, so a small number of parents might have fudged their offspring's date, time, and place of birth to give their little darling the most desirable natal chart possible. That skewing of the data—albeit from well-meaning and supportive families— might have been enough to give Gauquelin his correlation.

[H1] Gauquelin excluded basketball players, because apparently they gave "the most disappointing results in the European sample."

YOU ANSWERED E

Why not try … **which Harry Potter character are you? An online quiz!**

If you chose this answer, you're definitely a Hufflepuff.

YOU ANSWERED F

Why not try … **convincing everyone that there's no such thing as a fixed personality?**

That's what psychologist Walter Mischel tried to do in 1968, anyway. Going against the flow of many of his colleagues, Mischel suggested that "personality" per se does not exist. People don't conform to their "type" all the time, but behave differently depending on the specifics of the situation and their perspective on it. If you're under stress, do you react differently to a situation compared to when you're nice and relaxed? Is your behavior one way with friends and family, and another way with strangers?

If your answer is a definitive "no" then you're in the wrong answer—try B instead.

Life would be much simpler if we could tick-box, bump-feel, inkblot, birth-chart, or Harry-Potter our way to the life plan that is guaranteed to suit our inner selves.

If you think that sounds too good to be true, don't worry. You're not the only one out there with that particular personality type.

HOW TO MAKE **2D** GLASSES

Look at that cute rabbit, don't you just want to eat it?! I bet you do actually, you filthy carnivore. This poor bunny here is prey to many animals and you can see it in his face. Literally. Prey animals tend to have eyes on the sides of their faces. This is to give them more coverage, to help them see attacks coming from all angles. Compare them to predator animals like humans, who tend to have their eyes pointing forward. Seems a bit silly by comparison. What's the point of having two eyes if you're just going to point them in the same direction?

Prey   Prey   Prey

Predator   Predator   Technically
                            a predator

It turns out that training both your eyes on the same scene gives predators one huge advantage—3D vision! If your next meal is something you have to pounce on, you'd better be really good at estimating how far away it is. But how does having two eyes pointing in the same direction help you do that?

Try this experiment—find a small object in the distance or a spot on the wall, something like that. Now close one eye and hold up your thumb. Try to cover the object with your thumb so you can't see it. If the spot you chose is too big, find a smaller spot or walk backward until the spot is small enough. Or find a bigger thumb. Actually, if you look closely at your thumb, you'll find it *is* bigger. That's how looking closely works. What I'm saying is, bring your thumb closer to your eye if you need to. I shouldn't need to hold your hand like this.

Now, without moving your thumb, close your open eye and open your closed eye. You should now be able to see the object you thought you'd covered up. That's because your eyes are in slightly different places on your head, so the view they get is also slightly different. Like these two images here, taken from two spots about the same distance apart as your eyes:

Your brain is able to pick out those differences and put together a 3D picture of the world with all that useful information about how far away things are. This is called stereopsis. If you want to know what it feels like, then look again at those two images, but this time cross your eyes until they line up over each other. You should see it pop into a 3D image. Or just, you know, look up from this book at the real world.

This isn't the only way we perceive depth but it's the dominant one, and it's how 3D movies work. They show a slightly different image to your left eye and your right eye and your brain does the rest. It's usually achieved with a special pair of glasses and a special projector. The projector sends two slightly different images using two different types of light. The lenses in the glasses then filter the light to make sure the right image goes to the right eye (and the left image goes to the left eye).

But how can you have different types of light? The trick is to polarize it. You might know that light travels as a wave, which is to say, there's some vibration going on and it's possible to restrict that vibration to a single direction. It's like when you waggle the end of a rope up and down, the waves travel along the rope in an up-and-down direction not a side-to-side direction.

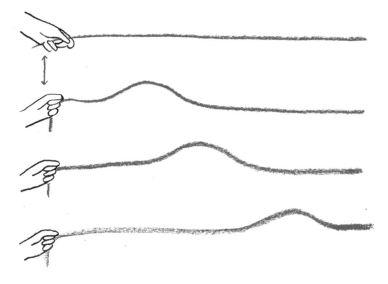

The lenses in 3D glasses only allow light through if it's waggling in the right direction.

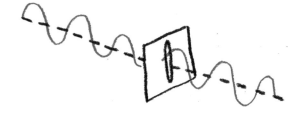

But not if it's waggling in the wrong direction.

If you take a pair of 3D glasses away with you from the movie theater, you can see the effect of the polarizing filter at home. Hold up your glasses to your flatscreen TV or laptop screen or iPhone screen (it doesn't always work on Android, depending on the model). Rotate the glasses and you'll notice the screen behind go light and dark as you turn them.[S1] That's because the light coming from LCD screens is polarized.

Some people really don't like 3D movies, which is why I'm going to show you how to make your own pair of 2D glasses. Next time you're dragged along to a screening with one more D than you'd like (I personally love the extra D), bring along these glasses and enjoy a nice flat viewing experience.

People who don't like 3D movies tend to complain about the way it makes them feel: headaches, sore eyes, dizziness, confusion. The main culprit appears to be the vergence-accommodation conflict. Whenever you look at something, your brain does two things in tandem. The first is to make sure both your eyes are pointing at the thing. That means, if it's close to you, your eyes will be pointing inward. Whereas, if it's far away, your eyes will be pointing almost straight out (see overleaf).

[S1] **If your glasses are from a circularly polarizing movie theater you may need to flip them over to get this to work.**

That's called the vergence, because your eyes are converging on something. At the same time, your brain makes sure the thing you're looking at is in focus. It does this by squishing and stretching the lenses inside your eyes until your vision isn't blurry anymore. It's working out the right focal length, basically, and this is called accommodation.

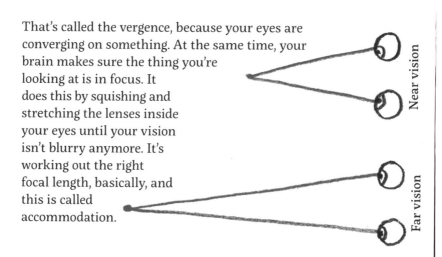

These two tasks go hand in hand. Which is to say, a particular vergence setting always corresponds to a particular accommodation setting. So when your eyes converge on an object your brain knows the corresponding focal length to set your lenses to. It's very clever and it all happens subconsciously.

Now, when you go to a 3D movie theater, the glasses trick your eyes into converging on something either closer or farther away than the screen really is. Your brain then sets your eyes to the matching accommodation, which is no good at all because for a clear image you need to focus on the screen itself.

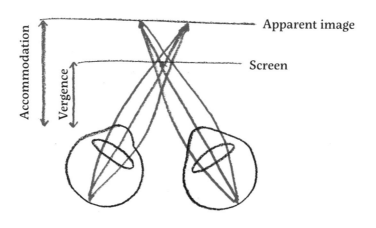

This all makes for a subconsciously unpleasant experience and is why you need your own pair of 2D glasses.

All you need for this project is two pairs of 3D glasses, and what you're trying to achieve is a pair of glasses with two left lenses (or two right lenses).

Most glasses given out in movie theaters are quite flimsy and you can pop out the lenses relatively easily. It's then simply a case of taking the right lens out and replacing it with the left lens from the second pair. Easy!

Now when you go to the movie theater, both your eyes will be served the same image, and if you have to pounce at the screen you'll know exactly how far to jump.

Monocular depth perception

We don't get our depth perception from stereopsis alone. Other techniques at our disposal include:

- **Accommodation**—how much we have to squish or stretch the lenses in our eyes to bring something into focus.
- **Motion parallax**—by moving our head around, even with one eye closed, we get a good sense of depth by observing how objects in our field of view move relative to each other.
- **Knowledge of the world**—if we expect lines in the real world to be parallel, we can use perspective to figure out depth. If we know how big an object is, we can then figure out its distance from us by how much space it takes up in our field of view.

THE VOICES INSIDE MY HEAD

Let's take a sneak peek at a conversation between the various different parts of my brain, recorded earlier today:

Helen 1 Hey, how's it going?

Helen 2 Fantastic! Couldn't be better.

Helen 3 Well …

Helen 1 I didn't ask you, Helen 3.

Helen 3 Sorry.

Helen 2 How are you doing, Helen 1?

Helen 1 Not great, if I'm honest.

Helen 2 What's wrong?

Helen 1 I'm writing this book, *The Element in the Room*.

Helen 2 No way! So am I! And it's going fantastically!

Helen 1 I mean, it's been going fine, but I'm really stuck on this chapter about brains.

Helen 2 How can you be? You've got one, haven't you?

Helen 1 I guess so …

Helen 3 I think you've got a lovely brain.

Helen 1 Shut up Helen 3, this isn't about you.

Helen 3 Sorry.

Helen 1 I just don't think that simply possessing something is what qualifies you to talk about it. I have a vinyl record of New Kids On The Block that I got for my twelfth birthday,[H1] but I'm not going to answer a pop quiz about it.[H2]

Helen 2 Hogwash. I've got a fantastic brain. Which means I know all about brains. Especially fantastic ones like mine. That overrides all other factors. What could possibly be more useful than first-hand experience?

Helen 1 Umm … well, how about a substantial body of research in the field? And a creative construct to get your ideas across in a witty yet illuminating manner?

[H1] This is true.
[H2] This is not true, I could totally get 100% with my epic NKOTB knowledge.

Helen 2	Forget that, I've got my own thoughts on the matter. I've read some blogs. I can bosh this whole chapter out in an hour, tops.
Helen 1	I admire your confidence, Helen 2.
Helen 3	I admire you both, Helen 1 and Helen 2.
Helen 1	Seriously, Helen 3. Haven't you got anywhere else you need to be?
Helen 3	Not really. Sorry.
Helen 1	Look, Helen 2. I'm not going to let this go so easily. How do you know you have such a great brain?
Helen 2	Stands to reason. I think, therefore I have a great brain. That's QED, duncehead.
Helen 1	This is so interesting. I've just been reading up on the Dunning–Kruger effect. I never thought I'd meet such an archetypal specimen, and inside my own head!
Helen 2	What's the Doodling Krunker effect?
Helen 1	It's this thing that psychologists David Dunning and Justin Kruger came up with in 1999.
Helen 2	Never heard of them. Obviously total idiots.
Helen 1	Well, they didn't so much come up with it as formalize it. It's been noticed by many before, from Confucius to Shakespeare to Darwin. It's the illusion of superiority caused by people's inability to recognize their own ineptitude. Which leads that person to believe they are more competent than they are.
Helen 2	Losers!
Helen 1	Exactly. It's a type of cognitive bias. If you survey a bunch of people about what they are good at, more than 50 percent will describe themselves as having above-average driving skills, sexual prowess, intelligence, and so on. They can't all be above average, can they?
Helen 2	I agree with that statement 110 percent.
Helen 1	Basically, it's when people are too incompetent to understand their own incompetence.
Helen 2	Sure, sure, interesting stuff but I'm definitely smart and not remotely incompetent. I have two degrees to prove it, so … jog on.

Helen 1 But it affects smart people too, when they're looking at something outside their specialism.

Helen 2 I don't see how that's possible. If I have a basic understanding of a topic, I can just extrapolate, right?

Helen 1 Indeed. A very basic understanding actually makes it even easier to fall victim to the DKE. Every problem seems so easy to solve when you have no idea about the nitty-gritty issues that an expert battles with on a daily basis. The devil is in the detail, as they say.

Helen 2 As *you* say. Not people like me. Details, shmetails!

Helen 1 It's like when a physicist pops into the medical research lab next door and is all like, "Hey, cancer scientists, you're doing it wrong! I've just thought of something that you appear to have missed in your decades of intense research conducted by hundreds and thousands of specialists on exactly this area of understanding. Let me show you what I mean. Here, hold my beer …"

Helen 2 Look, that was just the one time.

Helen 1 And then they go ahead and describe a totally wacky theory that was tested and rejected years ago, if only they'd read the literature or, you know, asked a biologist or something. When you start looking for it, you'll see the Dunning-Kruger effect all the time, even in yourself. Flicking through a newspaper, you'll read something about your own specialist subject—say, New Kids On The Block—and you can critically assess whether you agree or disagree with the writer's hypothesis. But watch out for your blind spot—for instance, an article about boyband 5ive from the same era— where you might be more easily swayed by false information. It's dangerous stuff in this new era of "alternative facts."

Helen 2 La la la la la la.

Helen 1 What are you doing?

Helen 2 La la la la I'm not listening la la la la.

Helen 3 I'm still listening, Helen 1.

Helen 1 Stop doing that, Helen 3. This isn't for you.

Helen 3 Sorry.

Helen 1 I dunno. Maybe I've got this all wrong.

Helen 3 What do you mean?

Helen 1 Still wasn't talking to you, Helen 3.

Helen 3 Sorry.

Helen 1 I've done extensive reading on this topic. I've read scientific papers. I've talked to psychologists. I've drafted and redrafted this section of the book a hundred times.

Helen 3 That's at least 97 times more than Steve …[S1]

Helen 1 But maybe I've got the wrong end of the stick? Maybe I shouldn't be talking about the Dunning–Kruger effect at all? Maybe everything I've ever thought is wrong?

Helen 2 I know what's happening. You've got imposter syndrome.

Helen 1 Oh. Of course. I should have seen this coming. That's so perceptive of you, Helen 2.

Helen 2 Really? I have no idea what imposter syndrome is. It sounds cool though, doesn't it? I've never heard it before. In fact, I think I just came up with it now. That phrase: imposter syndrome. Immmmmmposter syndrooooome! It's good, isn't it? Very proud of that one.

Helen 1 I can't believe you just Dunning–Kruger-ed the imposter syndrome.

Helen 2 What?

Helen 1 Imposter syndrome is a real thing. It's been a named phenomenon since even earlier than Dunning–Kruger, from 1978 when it was observed and written about by clinical psychologists Pauline R. Clance and Suzanne Imes.

Helen 2 More losers.

Helen 1 It's when a high-achieving person lives in constant fear of being exposed as a "fraud"

[S1] True.

	and thinks they don't deserve their success.
Helen 2	They don't. Hey, if you don't want your success I'll take it off your hands for free.
Helen 1	In extreme cases, they view their success as due to some external factor like luck or timing, even when evidence to the contrary is right in front of them. And the most ironic thing? Sometimes they put their success down to their own deceit, believing that they've convinced other people that they're smarter than they really are.
Helen 2	Just sounds like modesty to me. Pathetic!
Helen 1	But mere modesty comes with a genuine understanding of one's skills, then a decision to play it down for whatever reason: social cohesion, privacy, guilt … imposter syndrome is when someone is too uncompetent to believe they are not incompetent. If you know what I mean.
Helen 2	Not a clue, mate.
Helen 1	So, anyway, that's what's up with me. Helen 2 is about to prove she is suffering from the Dunning-Kruger effect …
Helen 2	No I'm not!
Helen 1	… exactly, and it turns out that I have imposter syndrome.
Helen 3	I love you.
Helen 1	How are you still here, Helen 3?
Helen 3	Sorry.
Helen 1	That reminds me, I haven't actually asked. What's up with you?
Helen 3	I've got Stockholm syndrome.
Helen 1	That explains a lot.
Helen 3	Can I go now?
Helen 1 and Helen 2	No.
Helen 3	I still love you both.

BRAIN MYTHS

You're a
left-brained person
or a
right-brained person

The corpus callosum is a band of neural fibers that connects the two hemispheres of the brain. In severe cases of epilepsy, where no other treatments have been effective, it can be cut to prevent seizures spreading from one side of the brain to the other.

Neuropsychologists Roger Sperry and Michael Gazzaniga discovered that in these "split-brain" patients, where the two sides can no longer communicate, the left and right hemispheres respond differently to stimuli, suggesting that they function differently and are used for different tasks. So far so scientific. The myth comes when pop psychologists tell us that our personalities are shaped by which side of our brains are dominant. So people who are details-focused—something your left brain is good at—should also be logical because that's handled by your left brain too. We simply don't see this in practice, however. Being good at one left-brain activity makes you no more likely to be good at another.

It might seem like a harmless myth, but I'd be nervous about my kids taking an online test to discover that they are right-brained and deciding that they won't be any good at math.

Are there *any* good personality tests out there? Don't forget to take our Personality Test Test (see page 80) to find out!

Men and women are just wired differently

Men are from Mars, women are from Venus. Or so we're told. Behavioral differences between the sexes are fascinating to both, and the common explanation is structural differences in the brain. But what does science have to say on the matter?

In a comprehensive study of existing data, researchers led by professor of psychology Daphna Joel analyzed the gray matter and white matter of 1,400 brains. They found various structural differences, like the left hippocampus was on average larger in men. But, crucially, there was a lot of overlap. They found that it was useful to talk about a continuum of maleness to femaleness and then see where different parts of the brains in the study fell on that spectrum. They found that most brains were a mosaic of typically female and typically male structures, with at most six percent of brains showing only male or female characteristics. So structurally, at least, there is no such thing as a male brain or a female brain.

Our behavior isn't just defined by the size of the bulk structures in our brains, though. It's defined by the way they are connected; our connectome.

Associate professor of radiology Ragini Verma and her team looked at nearly 1,000 individuals and found reliable differences in their connectomes. For example, women reliably have stronger pathways in regions associated with memory and social cognition.

What the study can't tell us is whether learning a "female skill" would lead to a more "female brain," or whether having a more "female brain" leads to more "female behavior," and likewise for the brains of men.

LOOK WHO'S TALKING?

Something has been nagging away at the back of my mind while we've both been writing this book. Barring the vanishingly small likelihood that we have written exactly the same number of words—one of us will have a higher word count.

If a range of popular claims are to be believed, it'll be me. According to one self-help manual, women speak a daily average of 20,000 words, while men manage just 7,000. Another claims 7,000 for women and 2,000 for men. What? It's like they just made these numbers up, instead of conducting robust, peer-reviewed studies!

So, do women talk more than men? Enter science, to crush stereotypes with statistics! Or support them! Let's find out.

A 2007 study led by James W. Pennebaker at the University of Texas recorded the chatter of nearly 400 male and female volunteers and came up a daily average for the number of words spoken by each group. And the women did talk more, by the smallest of margins: an average of 16,215 words compared to 15,669. If you overlaid those stats onto this book, I would have written just four more pages than Steve, out of 224. Less than two percent difference! That's what I call statistically insignificant.

Another nail in the coffin of this trope comes from the outliers: the volunteers who said the least (just 500 words) and the most (47,000 words) were both male. The study itself is no outlier either. Pennebaker had been looking at the question for a decade and seen the difference between the sexes was minimal to nonexistent.

But what about this book? The results are in:
      Helen: 24,869.   **Steve:** 19,820. [S1] **Matt:** 403.

So, out of a total of 45,092 words, the men wrote 45 percent of them and the woman wrote 55 percent. Of course this is just one sample ... we'll have to write a whole series of books to get any meaningful statistics. Back to the writing laboratory!

[S1] 19,821![S2]
[S2] 19,822![S3]
[S3] 19,823![H1]
[H1] OK Steve, we get it ...

ELEMENT STUFF

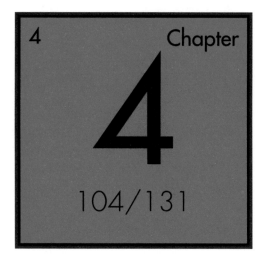

4   Chapter

4

104/131

So. Are we going to talk about the element in the room?

Let's quickly kill this joke by explaining it … This is the part of our book where we investigate a few elements of the periodic table that you might not expect to find in your house. The mercury in your lightbulb, the americium in your smoke detector, the potassium in your fruit bowl … and one that controls our daily lives from a distance: the tick-tick-tocking of cesium in clocks. And in all this elemental excitement, we mustn't forget the also-rans: element names that were mooted, but never confirmed, in the infamous Transfermium Wars.

Let's start with the elements that can always be found in every room you go into … because they are inside *you*!

THE ELEMENTS IN YOU

At a fundamental level, we are all made of stars.

That's right, the elements in our bodies were created by the deaths of countless giant suns as they turned supernova and exploded, scattering their matter across the Universe up to 12 billion years ago. The hydrogen in humans has been around even longer: 13.8 billion years, since shortly after the Big Bang.

It's an inspiring image. But we're going to take our heads out of the sky and bring them back down to earth because this chapter is about the elements in and around us: if it's not an element in the room, we're not interested![H1]

So here's a question about elements, and also about rooms.

If you took all the stardust, I mean elements, inside a human and laid them out on your kitchen table, what would it look like?

Please note that we do not recommend trying this at home ...

DO NOT TRY THIS AT HOME

The definitive list of elements that are essential, or at least important, to human life is still a matter of debate. Answers vary between 19 and 29, and sometimes include traces of boron, cadmium, chromium, cobalt, copper, fluorine, iodine, manganese, molybdenum, selenium, silicon, tin, vanadium, and zinc. None of these lists include the element of surprise.

But let's just take a look at the top 12 elements in your body.

[H1] Until we get to Chapter 6, which is all about space. Then we're interested. Very interested ...

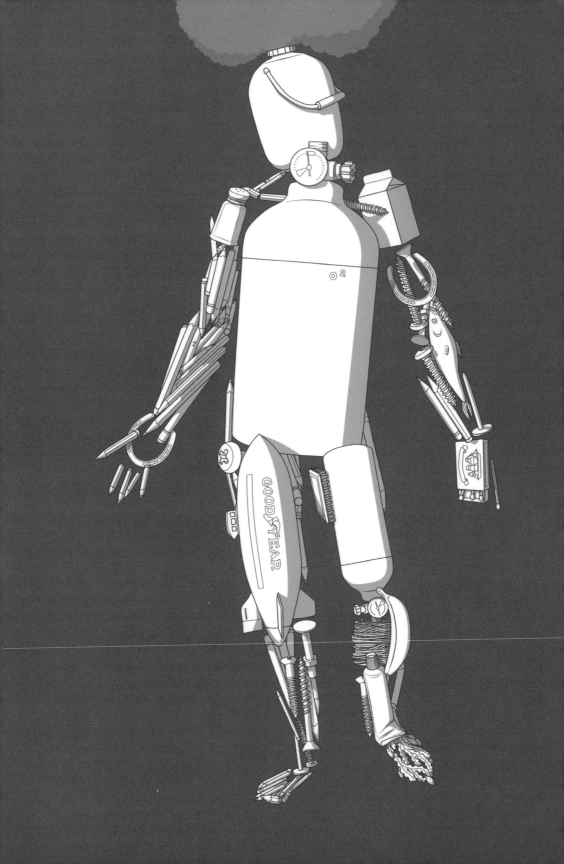

What are you *packing inside your person?*

ELEMENT	% OF BODY MASS
OXYGEN	65
CARBON	18.5
HYDROGEN	9.5
NITROGEN	3.2
CALCIUM	1.5
PHOSPHORUS	1.0
POTASSIUM	0.4
SODIUM	0.2
CHLORINE	0.2
MAGNESIUM	0.1
SULFUR	0.04
IRON	0.008

There's an issue here already. Three of the top four elements, nearly 80 percent of your mass, all naturally occur in gas form: oxygen, hydrogen, and nitrogen. Assuming your kitchen is at room temperature (70°F), a pressure of 1 atmosphere and you are of average weight (180lb), those elements will take up 1,400ft^3, 3,200ft^3 and 78ft^3 respectively.

If your kitchen is the size of a London double-decker bus, you'll just about squeeze it all in. Just don't turn on the stove, as the potent mix of hydrogen and oxygen will make everything a little too toasty at the slightest hint of a naked flame.[H1]

Moving on to carbon. With about 33 pounds of this fantastically diverse element in the average human, there are plenty of ways we can get that amount onto the table. Stacking up 12,333 pencils. Finding a 74,000-carat diamond. Spreading out a layer of carbon just one atom deep, to form a sheet of wonder-material graphene.

[H1] You could save space by combining two hydrogens with every oxygen and storing it as dihydrogen monoxide, AKA water. You'll only need to hold 16 gallons of that, which you can easily do in a box measuring 33½ inches in all directions, or in the fuel tank of a Dodge Caravan. It's much more compact, but this chapter isn't about compounds is it?

A sheet that is hundreds of times stronger than steel, conducts electricity faster than anything else on Earth, and covers an area six times the size of Central Park. Hmm. (Maybe it's time to plan that kitchen extension after all ...)

Then, it's calcium. An average person is hauling around about 2 pounds 10 ounces of this metal, with 99 percent locked into your bones and teeth. It's difficult to leave it in a single lump on the table, as it loves to react with other elements to make everything from limestone rock (calcium carbonate) to old-school rocket fuel (calcium permanganate). It's not terrifically interesting on its own, to be honest. Maybe you have a friend who is a serial monogamist? You know ... they do anything to avoid being single. Even if their pairings end up being incredibly boring, or highly explosive? Well, that's calcium.

Next up, some more metals. Imagine you have in your hands a lump of phosphorus the same weight as a loaf of bread. Pile onto that a block of potassium, about as heavy as a can of soup. Top off this tower with a chunk of sodium weighing the same as a billiard ball. Now imagine the blinding light as the air around you ignites the phosphorus in your hands, and imagine the searing pain as the water in your skin sets the potassium and sodium alight. Best throw it all out of the window quick, before the flames reach that stack of pencils ...

There's not much respite from disaster as you head down the rest of the list, with toxic chlorine gas, flammable magnesium, and blue-burning sulfur melting through your table into a pool of blood-red liquid on the floor.

Until you reach iron. Ah iron. Our old and stable friend. There's only just over 6.5 grams of iron in the average human, just over the weight of a quarter.

So, as your double-decker kitchen fills with noxious fumes and your refrigerator melts into a gooey puddle, hold your tiny disk of iron tightly in your fist and celebrate how great it is that humans come as prepackaged meat puppets already, and not as do-it-yourself stardust kits.

ONE BANANA
TWO BANANA
THREE BANANA
FOUR

Let's talk about one of the greatest measuring tools of all time. It's not the yard rule, the pound weight, the light year …

No, it's the banana.

More specifically, the Banana Equivalent Dose—or BED. It's a nice informal way of measuring radiation exposure and it's one that you can reach out and grab for yourself from the nearest fruit bowl.[H1] Because it turns out your average banana contains about 500mg of naturally occurring potassium, which is slightly radioactive.

Don't freak out! We're all getting tiny bits of radiation all the time, through cosmic rays from outer space, from the ground under our houses, from just going about our daily lives. And that's why the BED exists: to give some sensible context to any B-movie-style scaremongering that happens whenever the topic of radiation comes up.

Well, I say "sensible," but this is a scientific unit measured entirely in bananas.[H2]

What the BED tells us is how many bananas you would have to eat to get the same effective dose of radiation in any given situation.

[H1] Or possibly "herb bowl." Bananas are definitely fruit, but also technically herbs, as they grow on herbaceous plants that do not contain woody tissue, rather than actual trees.
[H2] Not officially. The BED is yet to be confirmed by the International System of Units. I'm working on it.

Why do physicists go bananas?

Just over 0.01 percent of the naturally occurring potassium on Earth is not like the rest—it's radioactive potassium-40. Instead of having 19 protons and 20 neutrons inside each nucleus like the more common potassium-39, this potassium-40 is an unstable version with 19 protons and 21 neutrons.

That's an extra neutron hanging out in the nucleus of just one in every 10,000 or so potassium atoms. When those "heavy" versions decay, they fling out highly energetic particles, making this everyday element ever so slightly radioactive.[S1]

What's so pleasing about using bananas is that the numbers work out beautifully: there's half a gram of potassium in your average banana, which generates a nice round amount of an *actual real* measure of radiation dose: 0.1 microsieverts, or 0.1μSv. But how much is a microsievert, and what does that amount of radiation mean for the human body? Well, let's use bendy yellow fruit to find out!

Before we kick off the comparisons, please don't stop eating bananas just because they're a bit radioactive. You need potassium to help regulate your blood sugar, keep your heart rate steady, and allow your nerve cells to function properly. And if you eat enough of them, you might turn into Bananaman.[H1]

Anyway, you get more radiation from a handful of Brazil nuts than a single banana,[H2] but years of experimental evidence in the field of comedy has proven beyond a shadow of a doubt that bananas are the funniest of all the foods. You can't hilariously slip up on the skin of a Brazil nut. And they look nothing like a penis.[H3]

[S1] For more about what really happens to radioactive potassium when it decays, see page 121.
[H1] You won't.
[H2] Not because Brazil nuts contain more potassium than bananas, but because they have the highly radioctive element radium in them. Bananas not looking so dangerous now, eh?
[H3] If you or someone you know has a penis that looks like a Brazil nut, please seek medical help immediately.

What's the equivalent of ...?

Average daily radiation exposure for an ordinary human just going about their business:

Up to 100 bananas   10μSv

Lying next to someone as you sleep

Half a banana   0.05μSv

Eating ¾ cup of Brazil nuts[H1]

100 bananas   10μSv

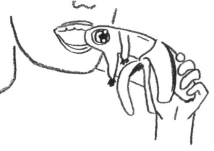

Eating a banana

1 banana   0.1μSv

Dental X-ray

50 bananas   5μSv

[H1] **Exposure levels provided by Public Health England.**

Day trip to Cornwall, UK[H1]

150 bananas 15μSv

Single chest X-ray

140 bananas 14μSv

Flying from London to New York

800 bananas 80μSv

This is where the Banana Equivalent Dose starts to break down. Eating 800 bananas in the time it takes to fly from London to New York is no mean feat. That's around 100 bananas an hour, not to mention what you'd have to do to get them through customs.

Living in the UK—average annual radiation dose

27,000 bananas 2.7mSv [H2]

[H1] **Due to radioactive radon gas released from the ground in this area.**
[H2] **1mSv, or one millisievert, is equivalent to 1,000 μSv or 10,000 bananas.**

Living in the USA—average annual radiation dose

62,000 bananas 6.2mSv

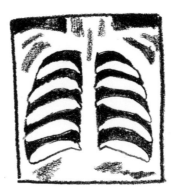

Hospital CT chest scan

66,000 bananas 6.6mSv

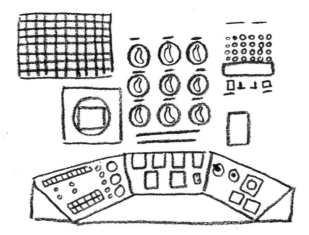

Average annual occupational exposure for nuclear power station worker (UK, 2010)

1,800 bananas 0.18mSv

Maximum annual limit for nuclear industry employees in the UK

200,000 bananas 20mSv

Exposure level that causes visible changes in blood cells

1,000,000 bananas 100mSv

Approximate number of bananas thrown away every day by UK households[H1]

1,400,000 bananas   140mSv

Radiation dose that would kill about half the people receiving it, within one month

50,000,000 bananas   5,000 mSv

OK, so for large radiation doses, the BED isn't very useful. Eating fifty-million bananas would cause a number of health problems, and death would be just one of many issues that you might encounter along the way.

After seven bananas you'd hit the UK limit for the recommended intake of potassium. After about 400 bananas the heart problems will kick in, as the balance of potassium in your cells goes haywire and the nerves and muscles that regulate your heartbeat can't function anymore. Although if you think about it, your heart won't be the first internal organ to feel the impact of 400 bananas ...

Also, potassium doesn't accumulate in the human body but is constantly "flushed out" to maintain your internal balance of different metal salts. There comes a point where the number of bananas you can physically eat over a short period of time is balanced out by your body's ability to get rid of the excess potassium. There's a natural plateau. Apparently. I mean, I haven't tried. But I'm prepared to take any biologist's word for it.

What I'm saying is, it might not be a great idea to use BED as the basis for your Health and Safety guidelines at work. Also, don't go putting radioactive stuff on your plate! Unless it's a banana.

[H1] According to 2017 estimates by Wrap, the UK government's waste advisory body.

MERCURY

You might have heard that mercury is poisonous, but are those fears all in your head? Well, if you've had amalgam fillings, then yes, and slowly leaching out into the rest of your body. But is mercury really dangerous? How much is too much? And what about other sources around the home?

Your first thought might be thermometers. But, actually, modern thermometers *don't* have mercury in them. If you've got an old one lying around, it might, but that's not how they're made these days. Which is a shame because mercury is perfect for the job:

PROS	CONS
It stays liquid over a wide range of temperatures.	Neurotoxin.
It doesn't wet glass, so when the temperature goes down, it doesn't stick to the inside of the tube and give a false reading.	
Its volume changes a lot with temperature, so it's really accurate.	
Its volume changes linearly with temperature, so a fixed change in height is always the same fixed change in temperature.	

OK, not quite perfect.

The only other widespread source of mercury in the home is compact fluorescent energy-saving lightbulbs.

Fluorescent lights have a tiny blob of mercury inside their tubes. Some of the blob turns into vapor and escapes. Before long, an equilibrium is reached where the amount of mercury escaping as vapor is the same as the amount of mercury vapor being reabsorbed back into the blob. When an electric current is passed through this vapor, the mercury atoms are excited into a higher energy state. When they settle back down, they release ultraviolet light. The coating of phosphor on the inner surface of the tube absorbs this light and reemits it as visible light.

PROS	CONS
Gives off ultraviolet light when excited.	Neurotoxin.

While the mercury in thermometers has been replaced with colored alcohol, there's no alternative for energy-saving bulbs. So do we need to worry about its toxic effects?

The expression "mad as a hatter" comes from the fact that hatters used a compound of mercury to smooth out felt and the fumes would cause "mad hatter syndrome." Symptoms include irritability and paranoia, similar to reading the comments section of the *Mail Online*. Mercury can even kill you. So why the hell do we put it in our teeth and light bulbs?

The UK Department of Health says that using mercury in fillings is free from risk. But let's put our tinfoil hats on for a second and do the math, just to be sure.

In a typical day, an amalgam filling could leach 2µg (micrograms) of mercury into your body. And if you break a compact fluorescent lightbulb and breathe in the delicious fumes, you'll ingest as much as 0.07µg of mercury.

How does that compare to something we all think of as perfectly safe: eating a tuna sandwich?

First of all, there's quite a bit of mercury in the ocean. Several industrial processes, such as burning coal, release it into the atmosphere and this newly freed mercury eventually winds up in the ocean via rainwater. When an animal at the bottom of the food chain such as zooplankton ingests that mercury, it stays in its body. And when the zooplankton gets eaten by a predator, the mercury stays in the predator's body. But because the predator has eaten loads of zooplankton, the mercury builds up. This is called bioaccumulation (which is, coincidentally, the name of my pet zooplankton) and it carries on up the food chain until you get to tuna, an apex predator at the top of the pile.

Tuna contains a decent amount of accumulated mercury, and when you eat a tuna sandwich, you're ingesting about 50µg of mercury. So to get the same exposure as eating one tuna sandwich a day, you'd need a filling in every tooth in your mouth. Or you'd need to break 700 lightbulbs. This level of exposure has no known health effects. But you might injure yourself bumping into things after you've broken all the lightbulbs to prove this very point.

And in a final weird twist, old-fashioned incandescent lightbulbs actually release more mercury into the environment than energy-saving bulbs. Even though the older bulbs have no mercury in them, they use more electricity, some of which is generated by burning coal, a process that releases mercury. The difference in energy consumption between the two types of bulb means the incandescent type is by far the worst offender.

All of this suggests that the mercury we have in our teeth and homes isn't so dangerous. Either that, or we're all already mad from the tuna sandwiches and we just don't know it.

Actually, thinking about it, people do say I'm paranoid.

Well, they don't say it but I know it's what they're thinking.

I LIKE TO BE AMERICIUM

Here at Festival of the Spoken Nerd, we've set off a few fire alarms in our time (see page 150—AKA Steve's spinning wastebasket fire).

You probably have one such detector on a ceiling in your home. If you don't, you really should ... and if you do, just be a good sport for us and give it a poke, check the battery still works, won't you?

At the heart of many smoke detectors[H1] is a radioactive element that has been forged in the furnace of a nuclear reactor Originally a by-product of the Manhattan Project, the secretive experiments of US scientists during the Second World War that ended in the development of the atomic bomb, it basically doesn't exist in the natural world. Discovered in 1944 and named americium after the Americas, it sits at No. 95 on the periodic table. That's directly below No. 63, the element named "europium" after the equivalent European continent.

Like americium, europium is also a rare beast, but it may have sneaked into your house already without your knowledge, via your wallet. It's used on Euro banknotes to color their anti-counterfeiting markings, which only reveal themselves under ultraviolet light. Nice. Subtle. Not like americium. Oh no.

This soft, silvery metal has one property that makes it ideal for safety equipment. That property, ironically, is the kind of radiation it constantly emits. Unlike other radioactive elements that are easier to extract or create, such as radium, americium lets off lots of alpha-particle radiation, but not much else.[H2]

Those alpha particles are basically the same thing as a flying helium nucleus—two neutrons and two protons stuck together —and because they're so heavy, they don't travel very far, and

[H1] There are several types of detector available. Some use ionizing radiation, as described in this chapter, but these are now considered too sensitive for kitchens. They may still be appropriate for other parts of your house, unless, of course, you regularly burn toast in your bathroom.
[H2] There's a bit of gamma radiation knocking around too, but not enough for it to be a health hazard.

are easily stopped by a piece of paper. That's great for keeping radiation to a minimum in your home. Although the casings of smoke detectors are usually made of something less flammable than paper. Just to be on the safe side.

Smoke Alarm Health and Safety Guidelines

BONK

Wheeee!

It's these bumbling alpha particles that help detect the tiniest of smoke particles. The radiation from a small lump of americium passes through an ionization chamber: a space inside the detector that is open to the air in the room. Because alpha particles carry electric charge, they create a small, constant electrical current in the air across the chamber. Anything that enters the ionization chamber—smoke particles, burned-toast particles, spray-on deodorant particles, a particularly well-directed sneeze—will disrupt the current's flow. That disruption is easily picked up by the detector, which triggers the alarm. Neat.

Fig. 1 Status of next-door neighbors: FRIENDLY

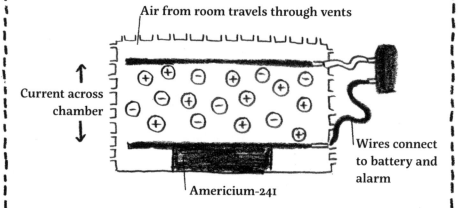

Air from room travels through vents

Current across chamber

Wires connect to battery and alarm

Americium-241

Fig. 2 Status of next-door neighbors: APOPLECTIC

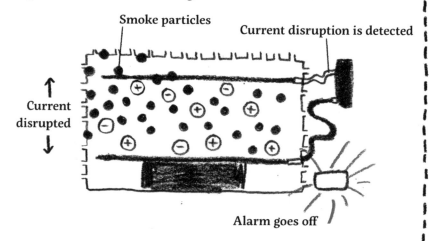

Smoke particles

Current disruption is detected

Current disrupted

Alarm goes off

Should you be worried about the tiny box of radioactive americium on your ceiling? Nope. Partly because not all smoke detectors use ionizing radiation—some use a beam of light that sounds the alarm when smoke particles get in its path. And partly because you'd get a higher radiation dose from sticking a banana up there, and that would be far less effective if your house catches fire.

ARGON AND THE HISTORY OF EARTH

Mercury isn't the only gas inside a fluorescent lightbulb. The other one is argon. In many ways the opposite of mercury, harmless and inert. And it has a fascinating story to tell.

Argon isn't like the other elements in the periodic table. It doesn't follow the rules. When you look closely at the masses of the elements and the pattern they follow, you see that argon sticks out like a sore thumb. The reason is that the Earth is very, very old.

The vast majority of the mass of an atom comes from its protons and its neutrons (electrons are superlight so we can ignore them). Protons and neutrons have the same mass, so we've come up with this handy unit of mass for the atom called the atomic mass. All you have to do to calculate the atomic mass is count up the number of protons and neutrons. For example, helium has two protons and two neutrons so its atomic mass is four (I'm simplifying very slightly, don't write in).

Typically atoms have one neutron for every proton but this rule isn't followed strictly—the obvious exception being hydrogen, which has one proton and no neutrons. But generally speaking the rule holds up well, up to about calcium on the periodic table.

Below is a plot of the number of protons and number of neutrons in the atoms as you go up the periodic table.

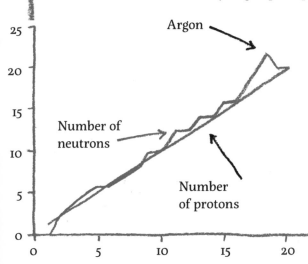

The first thing you notice is argon poking out above the rest. So much so that argon is actually heavier than the next element up, potassium.

Look more closely and you'll notice something else a little strange.

The number of neutrons isn't always a whole number. How can that be? Do atoms have fractions of neutrons in them? The answer is much more mundane. Elements can have different numbers of neutrons so we take the average of what we see around us. For example, carbon usually has 6 neutrons but is occasionally seen with 7 or 8 neutrons, which bumps the average up slightly to 6.01.

The funny thing is, argon *used* to follow the rule quite well, and elsewhere in the Universe it still does. The blip is only here and now on Earth. But why?

It's all to do with how argon is made, and the method of production is different on Earth than it is in the rest of the Universe. Most of the argon in the Universe is made by fusion inside stars, and this method overwhelmingly produces argon with the same number of protons and neutrons: 18 of each. But on Earth, most of the argon comes from radioactive ~~bananas~~[H1] potassium decaying into argon. This method produces a heavier type of argon with 18 protons and 22 neutrons.

So the Earth has a tiny amount of sun-born argon in its atmosphere but way more of the potassium-born stuff. In fact, potassium decay has produced so much argon it's now the third most abundant element in the air we breathe.

It was a slow process, though. Radioactive elements decay at different rates and potassium is particularly slow. In fact, it would take about a billion years for half the atoms in a lump of radioactive potassium to decay (and only 10 percent of that would decay into argon, the rest would turn into calcium). So argon is only as heavy as it is now because radioactive potassium has been decaying for the whole of Earth's 4.5-billion-year history.

When Russian chemist Dmitri Mendeleev put together the periodic table in the 1860s, he began by arranging the elements by mass and then looking for repeating patterns in their behavior. In many cases the masses of the elements were crude estimates,

[H1] See page 109 for further details.

so if Mendeleev saw a pattern that required a heavier element to go before a lighter element, he would allow it on the assumption that the masses were wrong. He couldn't have known that in the case of argon and potassium the masses were right, and that this funny little blip shines a light on Earth's ancient history.

CESIUM THE DAY

This element is stretching the definition of "in the room," because it isn't actually in the room with you. But without it you couldn't do a lot of the things you do inside rooms, like watch TV, make cell phone calls, browse the internet, and more.

All the tech we take for granted today needs the answer to one very important question for it to operate. It doesn't just need the answer to be correct. It needs to be exactly correct absolutely everywhere on Earth, and in outer space.

That question is: how long is one second?

It's more difficult to answer than you might first imagine.[S1]

Gone are the days when you could measure the length of one rotation of the Earth around its axis, divide that by 24, by 60 and by 60 again, and get the length of a single second. Although that is actually a pretty good definition. If you measured the time between exactly midday one day and exactly midday the next, averaged it out over a few hundred days and used that "mean solar day" to calculate one second, you'd be out by maybe a thousandth of a millisecond, tops.

But imagine if the average time it takes for the Earth to rotate ever changes in the future? It would throw that calculation totally out of whack. This isn't just a theoretical issue: the mean solar day is gradually getting longer, thanks to interactions between the Moon, the Earth, and the tides. Not by much, but

[S1] Although a commonly accepted definition of a second is the length of time between sending a tweet with a spelling mistake and someone pointing it out.

enough to look for another solution. It needs something more reliable, more fundamental, more ... elemental.

Things running on different definitions of time cause problems. Delays of just a microsecond on stock-exchange communication can affect transactions to the tune of millions of dollars. GPS relies on absolute time coordination across every terrestrial device and extraterrestrial satellite or we're all, literally, lost.

Even in Britain in the 1840s, people knew that having the same time mattered, as the new national railroad network began to run cross-country services. Until then, it hadn't been a problem that each city would take their time from local sundials. The city of Oxford ran five minutes behind Greenwich Mean Time. Norwich, to the east rather than west of London, ran a few minutes ahead. Bristol clocks chimed the hour a full ten minutes behind the capital. As a network of tracks and trains built up across the country, it became obvious that using different times in different cities ran the risk of timetabling "snafus" or, much worse, actual rail accidents.
So it was decided that everyone would follow London Time. Except Bristol, who held out until 1852 by adding a second minute-hand to their clocks to show official "Railroad Time" as well as their own local Bristol Time.

In comes a role for the element that is in every room by proxy ... the cesium clock!

Though the phrase "cesium clock" is not a great description of what is involved. A volatile metal with similar properties to sodium and lithium, cesium explodes on contact with water. You could make an actual clock case, winding mechanism, and gears out of cesium, but at the very least I'd call that unadvisable.

Instead, it's over to the scientists of the National Physical Laboratory (NPL) in Teddington, London, who created the first accurate cesium clocks in 1955. They still run the UK's cesium

Time Lord

National Physical Laboratory

Hampton Road

Teddington

clocks today. They're responsible for defining our sense of time on a national scale, and are literally known as Time Lords. Coolest business card, ever.

Since 1967, one second has been defined as "the duration of 9,192,631,770 periods of the radiation corresponding to the transition between the two hyperfine levels of the ground state of the cesium-133 atom."

Translated from the Gallifreyan, that means using a bunch of high-powered lasers to cool down a blob of cesium to a fraction of a degree above absolute zero—so it hovers just above the coldest possible temperature in the Universe.

Then another bunch of lasers kick some of those cesium atoms up into the air and let them fall—a little bit like a fountain, but confusingly the fountain is made from atoms of supercold metal and does not involve any water.[H1]

Our brave Time Lords then blast the falling atoms with microwaves to make them flip between two different energy states. It turns out that the timing of the cesium flips can be measured to an incredible degree of accuracy. NPL's clocks are able to tell the time without losing or gaining more than a single second in 158 million years. It doesn't stop there ... the next generation of atomic clocks use even more accurately measurable elements like strontium and ytterbium. Flipping between the energy states of those metals will keep time to within one second in 14 billion years. That's longer than the Universe has been around. Damn, those Time Lords are good.

[H1] Thankfully. See previous information about the properties of cesium in water.

There is a moral to this story: your short time on Earth is being measured with the kind of accuracy that our simple humanoid brains can barely comprehend.

So don't waste it:
Carpe your atomic clock.
Cesium the day!

THE **-iums** THAT GOT AWAY

On November 28, 2016, I finally had to face the fact that my life's greatest ambition would never be fulfilled. The four most recently discovered elements—until that day, not yet officially named—were finally given their official titles, and none of them were called HelenArnium. There wasn't even a cheeky SteveMouldon either.

Instead, IUPAC[H1] finally agreed to name three after the regions of their discovering: nihonium (Nh) for Japan, moscovium (Mc) for Moscow, tennessine (Ts) for Tennessee, and one for the father of synthetic element creation Yuri Oganessian in Oganesson (Og).

Fine! I didn't want any of your cruddy new elements with my name on anyway!

OK, so cruddy is the wrong word, but actually I think it's for the best. Some of those elements have only been seen as a handful of atoms. Some last for less than a fraction of a second. None of them have any useful practical applications whatsoever. On second thoughts, you can keep them, Yuri and co.

Instead, I'll settle for spelling my name out in existing elements from the periodic table.

[H1] That's the International Union of Pure and Applied Chemistry. They get to choose the element names. I mean, I'm sure they do lots of other stuff, but as far as I'm concerned that's their main role.

PERIODIC TABLE OF THE ELEMENTS

Legend:
Atomic number → 1
Symbol → H
Name → Hydrogen
Atomic weight → 1.008

1 IA	2 IIA	3 IIIB	4 IVB	5 VB	6 VIB	7 VIIB	8 VIIIB	9 VIIIB	10 VIIIB	11 IB	12 IIB	13 IIIA	14 IVA	15 VA	16 VIA	17 VIIA	18 VIIIA
1 H Hydrogen 1.008																	2 He Helium 4.002602
3 Li Lithium 6.94	4 Be Beryllium 9.0121831											5 B Boron 10.81	6 C Carbon 12.011	7 N Nitrogen 14.007	8 O Oxygen 15.999	9 F Fluorine 18.998403163	10 Ne Neon 20.1797
11 Na Sodium 22.98976928	12 Mg Magnesium 24.305											13 Al Aluminium 26.9815385	14 Si Silicon 28.085	15 P Phosphorus 30.973761998	16 S Sulfur 32.06	17 Cl Chlorine 35.45	18 Ar Argon 39.948
19 K Potassium 39.0983	20 Ca Calcium 40.078	21 Sc Scandium 44.955908	22 Ti Titanium 47.867	23 V Vanadium 50.9415	24 Cr Chromium 51.9961	25 Mn Manganese 54.938044	26 Fe Iron 55.845	27 Co Cobalt 58.933194	28 Ni Nickel 58.6934	29 Cu Copper 63.546	30 Zn Zinc 65.38	31 Ga Gallium 69.723	32 Ge Germanium 72.630	33 As Arsenic 74.921595	34 Se Selenium 78.971	35 Br Bromine 79.904	36 Kr Krypton 83.798
37 Rb Rubidium 85.4678	38 Sr Strontium 87.62	39 Y Yttrium 88.90584	40 Zr Zirconium 91.224	41 Nb Niobium 92.90637	42 Mo Molybdenum 95.95	43 Tc Technetium (98)	44 Ru Ruthenium 101.07	45 Rh Rhodium 102.90550	46 Pd Palladium 106.42	47 Ag Silver 107.8682	48 Cd Cadmium 112.414	49 In Indium 114.818	50 Sn Tin 118.710	51 Sb Antimony 121.760	52 Te Tellurium 127.60	53 I Iodine 126.90447	54 Xe Xenon 131.293
55 Cs Caesium 132.90545196	56 Ba Barium 137.327	57 - 71 Lanthenoids	72 Hf Hafnium 178.49	73 Ta Tantalum 180.94788	74 W Tungsten 183.84	75 Re Rhenium 186.207	76 Os Osmium 190.23	77 Ir Iridium 192.217	78 Pt Platinum 195.084	79 Au Gold 196.966569	80 Hg Mercury 200.592	81 Tl Thallium 204.38	82 Pb Lead 207.2	83 Bi Bismuth 208.98040	84 Po Polonium (209)	85 At Astatine (210)	86 Rn Radon (222)
87 Fr Francium (223)	88 Ra Radium (226)	89 - 103 Actinoids	104 Rf Rutherfordium (267)	105 Db Dubnium (268)	106 Sg Seaborgium (269)	107 Bh Bohrium (270)	108 Hs Hassium (269)	109 Mt Meitnerium (278)	110 Ds Darmstadtium (281)	111 Rg Roentgenium (282)	112 Cn Copernicium (285)	113 Nh Nihonium (286)	114 Fl Flerovium (289)	115 Mc Moscovium (289)	116 Lv Livermorium (293)	117 Ts Tennessine (294)	118 Og Oganesson (294)

Lanthenoids (57):

57	58	59	60	61	62	63	64	65	66	67	68	69	70	71
La Lanthanum 138.90547	Ce Cerium 140.116	Pr Praseodymium 140.90766	Nd Neodymium 144.242	Pm Promethium (145)	Sm Samarium 150.36	Eu Europium 151.964	Gd Gadolinium 157.25	Tb Terbium 158.92535	Dy Dysprosium 162.500	Ho Holmium 164.93033	Er Erbium 167.259	Tm Thulium 168.93422	Yb Ytterbium 173.045	Lu Lutetium 174.9668

Actinoids (89):

89	90	91	92	93	94	95	96	97	98	99	100	101	102	103
Ac Actinium (227)	Th Thorium 232.0377	Pa Protactinium 231.03588	U Uranium 238.02891	Np Neptunium (237)	Pu Plutonium (244)	Am Americium (243)	Cm Curium (247)	Bk Berkelium (247)	Cf Californium (251)	Es Einsteinium (252)	Fm Fermium (257)	Md Mendelevium (258)	No Nobelium (259)	Lr Lawrencium (266)

Ah.

Dear Mendeleev,

we have a problem.

Hen Arney

Goddamit, IUPAC! Sort out your alphabet! How am I going to get that printed onto a shower curtain?

At least it's not as bad as Steve's version:

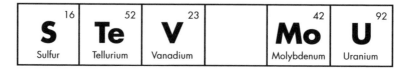

And what the hell is this?

26 Fe Iron	16 S Sulfur	22 Ti Titanium	23 V Vanadium	13 Al Aluminium	8 O Oxygen	9 F Fluorine	90 Th Tellurium	16 S Sulfur	84 Po Polonium	19 K Potassium	7 N Nitrogen	Ne Neon

Here's the thing. There are 118 elements on the periodic table, which we all know and love, and sometimes even sing about.[H1]

But some of those 118 names could have been quite different. It took the IUPAC about 12 years to recognize the four new elements as the real deal from the first undisputed sighting of nihonium in 2004. Then it took another year to agree the actual names, officially suggested by the teams who created

[H1] Check out my YouTube channel for the full version of Tom Lehrer's "The Elements" complete with all the new ones, chemistry nerds!

them, and unofficially suggested by absolutely everyone else on the planet.[H1]

Naming the latest four was actually relatively easy, compared to the hullabaloo of the element-naming business from the 1960s to the early 1990s. It was during the Transfermium Wars, so called because they involved the elements from fermium (number 100) onward. The Cold War meant that scientists across the world—especially in the nuclear physics laboratories of West Germany, Russia, and the USA—were all working in parallel rather than together, in a race to discover new elements that might possibly be used to blow each other up in ever more inventive ways. OK, so they were also looking for other uses for these new elements, but military research provided a lot of funding so top billing goes to them.

Here's what happened during those wilderness years for element-naming, in no particular order:

Clears throat, takes deep breath

The Americans pitched **seaborgium** (Sg) as element 106, after chemist Glenn T Seaborg. Unfortunately, that was a no-no because it broke with an important tradition: Seaborg was still alive.[H2]

Moving on. Rival proposals saw Russia naming element 105 **nielsbohrium** (Ns) after Danish physicist Niels Bohr, and the USA calling it **hahnium** (Ha) after German chemist Otto Hahn, whose name was also suggested for element 108, although that ended up as the similar-sounding **hassium** (Hs) after the German state of Hesse.

Darn it. I was so close to getting my initials on an element! There's more to come though ... Still with me?

As a compromise, IUPAC suggested **joliotium** (Jl) for 105, after Marie Curie's daughter and son-in-law team of element-hunters, the Joliot-Curies. None of those ideas made it to the final, and element 105 became **dubnium** (Db), after the Russian research center at Dubna.

[H1] My proposal for HelenArnium (Ha) never stood a chance, but an online petition demanded that one of the new elements be named "Lemmium," after Motörhead's frontman Ian "Lemmy" Kilmister, who sadly passed away just as the new discoveries were officially recognized. It made sense, Lemmy was all about the heavy metal. There were 157,185 signatures on the petition last time I checked, but sadly it never swayed the IUPAC.

[H2] Hmm ... having an element named after me is looking a lot less attractive, if that's the kind of hoop I have to jump through.

At one point, dubnium was also suggested for element 104, but the Russians wanted that one named **kurchatovium** (Ku) after their man Igor Kurchatov.

Meanwhile, **rutherfordium** (Rf) was the Russian favorite for element 103, but the same name was American frontrunner for element 104. IUPAC instead suggested naming the totally different element 106 after Ernest Rutherford, the New Zealand-born British physicist who first outlined the structure of the atom.

In the horse-trading that followed, 104 got rutherfordium, and 103 became **lawrencium**, or Lr, formerly known as Lw, named after Ernest Lawrence, the American inventor of the cyclotron where many of these synthetic elements were discovered. Element 106 eventually settled for seaborgium, even though Glenn was very much still alive and kicking.

Have you got all that?

Element 102 was suggested in turn as **nobelium** (No), **joliotium** and **flerovium** (Fo). In the end the "Noes" had it—as in, it ended up nobelium, as in "No," after all. Come on, keep up!

After that, some bright spark saved up **flerovium**, named after the Russian Flerov Laboratory, and used it for element 114, but that's a whole other can of worms we don't have time to open up here.

Otto Hahn's hahnium faced the jury yet again in the running for element 110, alongside French physicist Henri Becquerel's **becquerelium** (Bc), but it was the city of Darmstadt's **darmstadtium** (Ds) that made it onto the final table.

Good old nielsbohrium from element 105 almost rode again as element 107, but lost half of itself (half the name, not half the element) to become **bohrium** (Bh).

I hope you're memorizing all this. There will be a test at the end ...

Probably my favorite of all the transfermium elements is number 109, which at one point was touted as the thrice-rejected hahnium. In the end, 109 became **meitnerium** (Mt), which named an element in honor of one of those few female nuclear physicists who were able to flourish in this male-dominated scientific field. Her name was Lise Meitner and the fact that her groundbreaking work in nuclear fission wasn't named as part of the 1944 Nobel Prize for Chemistry is a travesty to this day.

So, who won the Nobel that year instead, you ask?

That's right: Otto Hahn. Him of almost-having-element-105-named-after-him fame. And almost-element-108, almost-element-109, and almost-element-110.

Hahn may have got his name on a gold medal, but Meitner got her name on the periodic table.

I'd call that one:nil to Meitner.

Perhaps in a parallel universe, hahnium won out and my initials are on a real actual element! But in that same world Lise Meitner gets nothing.[S1] I think I'll stick with this reality.

Lise Meitner, nuclear physicist and professional badass.

[S1] Not sure if it's helpful to point out that I already have my initials on the real periodic table, in samarium (Sm). No? Not helpful? OK, cool. Surely the rejected name bastardium deserves a mention though? No? Really? You bastardium ...

EXPERIMENT STUFF

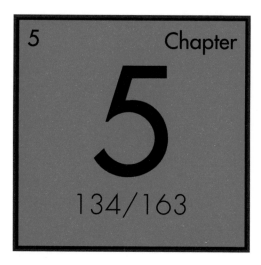

5   Chapter

5

134/163

Even the most tepid soirée can be rescued with the judicious addition of science. Follow the simple step-by-step directions in this chapter to get you and your guests experimenting all night long. The only things you'll need are bits and pieces you already have lying around the house and an irrepressible sense of derring-do.

We nerds aren't very selective about our shindigs and even the opening of an envelope can become an experimental happening that results in a special glow under the covers. We'll also show you how to enhance the mood lighting with spinning fire and rings of smoke, let go of your inhibitions with science cocktails, and accurately measure the buzz in the room with a randomized controlled trial.

But, first let's party like it's 1799, with sparks and shocks, just like the Georgians did.

MY LITTLE PHYSICS EXPERIMENT

People often ask me how I got into science. A little embarrassed, I tell them it was thanks to everyone's favorite anatomically distorted plastic horse toy: My Little Pony.

I'll explain. One of my most prized possessions as a seven-year-old child was a lurid pink My Little Pony nightdress, made from 100 percent purest polyester. Don't judge, this was the 1980s. Everything was made from polyester back then.

This nightdress was a fashion statement that makes the adult me look back on in horror. It was a fashion statement that made the eight-year-old me look back on in horror as well. But none of that mattered at the time because this nightdress was more than just an item of clothing. It was equipment … for science!

Being a young and curious nerd, I noticed something weird when I snuggled up under the covers at night. Whenever I wore this horse-based sartorial monstrosity, even the slightest movement under the darkness of a duvet would create tiny flashes of light and delicate crackling noises. Obviously I didn't know at the time that this was physics, I just thought the flashes were part of the magic of My Little Pony.

But no, it was static electricity. That thing that sometimes gives you a tiny shock when you touch a car door; that thing that makes a balloon stick to a cat with hilarious consequences; that thing where electrons are separated from their orbitals and collect together on the surface of an object until something, either accidentally or on purpose, releases them back to the earth. That thing that most people hardly think of as science any more, it's so ordinary and everyday.

But for the seven-year-old me, discovering static electricity was a formative experience. It set me on a path of curiosity that led to science, and I'd like to thank My Little Pony for that. Despite their use of sweaty polyester in a range of children's nightwear, they still managed to create the perfect equipment for making electricity under the covers. Although later in life I found out that it had the opposite effect in an actual relationship ...

Static electricity parties of yesteryear

These early experiments of mine were not the first highly charged gathering by any means. In the 1700s electricity parties were all the rage. Some used homemade contraptions that rubbed glass balls against woollen cloths to generate a supply of static electricity, which could be used to perform astounding tricks such as ... turning the pages of a book without anyone touching them, or making fake spiders dance in the air ... ooooh ...

Other more extravagant events involved a live electric eel. After the guests joined hands, the unluckiest partygoer would dip their fingers into the eel's tank to electrify them all. This may have been the origin of that party favorite, the "conger line."[S1]

[S1] It wasn't. And it's "conga line" anyway.

How many monks does it take to change a lightbulb?

About 200,[H1] according to a 1746 experiment. Although they weren't actually changing a lightbulb, but instead being used to create a primitive human power cable. Jean-Antoine Nollet, part-time scientist and full-time Abbot at the Carthusian monastery in Paris, persuaded a couple of hundred of his brothers to hold a 26-foot iron wire in each hand. They should have seen trouble coming at that point but still they went ahead, linking up with one wire between each of them to create a line more than a mile long.

Nollet released a powerful static shock into the first monk. It was recorded that the convulsions and yelps of pain traveled along the entire length of the line almost instantaneously. The monks proved that electricity is capable of transporting a message across long distances, faster than a human could carry it, which paved the way for the electric telegraph of the future. Unfortunately, the only message Nollet could send at the time was "Ow!"

The Body Electric

Electricity parties continued well into the 1800s, and as technology improved, the reliance on eel-power and monk-wires faded out and were replaced by more reliable methods of storing and transferring static electricity. John Cuthbertson, a British scientific instrument-maker, published *Practical Electricity and Galvanism* in 1821, to boost sales of his electricity-generating build-at-home science kits. Alongside instructions for how to "draw the electric aura from a patient"[H2] there were a number of

---

[H1] Some accounts suggest that 700 monks were involved, but the monk-math just doesn't add up.

[H2] In the early 1820s, electricity was seen as the cure for almost any ailment, from gout to blindness to reviving a person after drowning. Even genitourinary infections were treated, though hopefully not by direct application to the affected area.

step-by-step guides to performing party tricks, including what became known as the "Electric Venus," in which a lady of the party would be charged up with static electricity while standing on an insulated stool. Gentlemen would be invited to kiss her, but as each suitor came in for their clinch, sparks would fly in all the wrong places and the chaps would be more likely to melt their moustache than melt into a kiss.

It's quite a tribute to the fashion of the day to know that something similar to a Van de Graaff generator could be completely hidden under a regular bustle and crinoline. Although you'd think someone would have noticed the humming noise ...?

How to host a static electricity party

So, what better experiment to conduct at a party than one that quite literally "conducts"?

Get a rubber balloon. Type and color not important. If you're having a party, you're bound to have some lying around already.

Get an energy-saving fluorescent lightbulb. Type quite important. Kitchen cabinet tubes and those spiraly lamp bulbs work best.

Turn off all the lights or hide under a thick duvet. As you're having a party, you are probably going to be doing this anyway, just for fun. I guess. If you go to the kind of parties I do. You might need to spend a bit of time in the dark for your eyes to adjust to the low light levels and be able to see the effects of the static experiment. Ten minutes should do it. Three or four days may be too long.

Rub the balloon vigorously on long clean hair, a fluffy sweater, a nylon carpet, a passing cat, or your childhood My Little Pony nightdress that you've kept in a secret box in the attic since you were seven. What you're doing here is trapping lots of bonus electrons on the surface of the balloon, and building up a negative static "charge." Because the rubber of the balloon is a great electrical insulator, the electrons are trapped and can't go anywhere ... at least for now ...

If you're a particularly pedantic partygoer, you may have noticed that nothing is happening. Inflate the balloon, then repeat Step 4.

Once the balloon is fully "charged" move it gently toward the middle of the tube on your fluorescent lightbulb.[H1] The electrons trapped on the balloon will rush across into the lightbulb, usually making a tiny crackle or pop sound. Anywhere you get a stream of moving electrons, you get an electrical current ... and that electrical current will make the bulb light up for a very brief moment.

Ta-daa! Your static electricity party has started!

STEP 1
STEP 2
STEP 3
STEP 4
STEP 5
STEP 6

[H1] Do not plug the lightbulb into the household power. That is definitely cheating.

YOU ARE CORDIALLY INVITED TO THE OPENING OF AN ENVELOPE

Grab some junk mail and go somewhere dark (like under your duvet again). Peel back the flap of the envelope a little and put your eye up against the gap. Now continue to peel the flap away. You should see a thin line of blue light as the glue yields.

It's not just envelope glue that does this. Crushing a Life Savers mint will have a similar effect.

This is called triboluminescence; the light that is produced when chemical bonds are broken.

> "Breaking chemical bonds can sometimes lead to charge separation. When the charges come back together, light is emitted. Or something." **Scientists**

The phenomenon isn't well understood. But it is suspected that the energy required to pull charged particles apart (the energy you put into the glue when you tear open an envelope) is converted into light when the charges recombine.

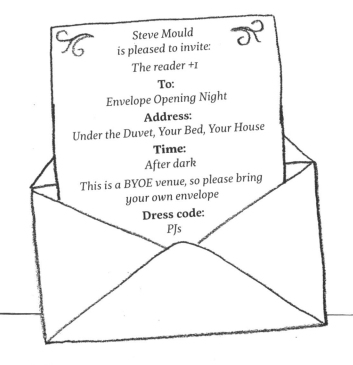

Steve Mould
is pleased to invite:
The reader +1
To:
Envelope Opening Night
Address:
Under the Duvet, Your Bed, Your House
Time:
After dark
This is a BYOE venue, so please bring
your own envelope
Dress code:
PJs

SELF-SIPHONING BEADS

Early in my career I was asked if I could do a chemistry show for a kids' party. My degree was physics, making me woefully unqualified. But I said yes, thinking that I'd figure out the details closer to the time (chemistry is just a small part of physics anyway, so I thought I should be fine).

The challenge when putting together a chemistry show is getting hold of all the fun chemicals such as liquid nitrogen. So instead I made the show all about polymers, which are basically just plastic and easy to get hold of. The show was full of weird and wonderful molecules such as polyethylene oxide, a really long molecule that when dissolved in water will pour itself out of a beaker!

Strange behavior that is crying out for an explanation. I'd seen Steve Spangler, an American science communicator, explain it with a chain of plastic beads in a beaker. The chain mimics the long molecule of polyethylene oxide—pull just a small bit out and the rest follows all on its own.

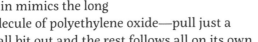

I wanted to see if I could reproduce this effect, but instead of plastic beads, I thought I'd get some metal ones. So I bought 160 feet of the type of chain that hangs down beside office blinds—the chain you pull on to open and close the slats.

Luckily, the effect works with metal beads too. But something else also happened that was remarkable, the chain rose above the pot before it fell back down! A strange sort of fountain effect.

I love a physics mystery; something odd that I can't explain. So I hopped on Google in search of the answer. And to my surprise, I couldn't find a single reference to the phenomenon (and my Googling skills are excellent). I was going to have to rely on my own actual brain, which I rarely do these days.

I really should have been able to figure this one out. It's just forces and vectors and conservation of energy and all that. Except I couldn't. Maybe I was out of practice.

So I decided to outsource the problem. I filmed the effect and uploaded the video to YouTube. My thinking was, if enough people see it, someone will eventually write an eloquent explanation in the comments. Which, it turns out, is not how YouTube comments work. That's not to say I didn't learn anything. For example, I learned that my experiment is "fake" and my face is "weird." Fascinating stuff.

The video was picked up by news aggregation website Reddit, pushing it to a million views in just a few days. But the lengthy discussion that followed there, and on other websites, was inconclusive.

If only I could film the thing in slow motion, I'd be able to pore over the footage and figure it out. Slow-motion cameras are expensive, though, so I put out a request for help. A YouTube channel called Earth Unplugged invited me to film it on their kit, and we captured some amazing shots.

I pored over the videos frame by frame and finally worked out: slow-motion footage doesn't help.

It seemed like the answer would elude me until I was contacted by two physicists at Cambridge University. They'd seen the slow-motion video and were compelled to put their sizable minds to it. John Biggins and Mark Warner wrote up their findings for the scientific journal, *Proceedings of the Royal Society A: Mathematical, Physical and Engineering Sciences* in a paper titled "Understanding the Chain Fountain."

The paper even has my name in the references, of which I'm immensely proud.

The answer?

Biggins and Warner considered the flexibility of the chain and how that affects its dynamics. They observed that the chain is very flexible but that if you try to bend it beyond a certain point, you can't; it becomes rigid. You can try this for yourself. Just grab some office-blind chain (or the chain on your dog tag) and pinch it. You form a tight little loop that won't go any tighter, like in the top corner of this page.

The dynamics of such a chain are quite complicated and hard to model mathematically. But I'm going to let you in on a secret about physicists: whenever the real world is too complicated, they just pretend it isn't! You might remember an example of this from school when you studied pendulums. Real pendulums are extremely complicated, but if you pretend there's no air resistance and that the string doesn't weigh anything, the math becomes much easier. And so it is with all of physics. The real world is too complicated so we imagine it isn't and then add complexity as we figure things out.

You can closely model the behavior of our bead chain with a simpler chain—a chain made of rigid sections joined by flexible links; essentially, a chain of rods.

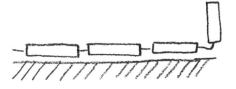

Now let's consider what happens when the chain is pulled up from one end, like when it is lifted from the pot. And let's focus on two rods in particular—the rod that has just been pulled into motion and the next rod down, which is at rest but about to be pulled out.

The rod at rest will feel an upward force from the right side. This is off center, meaning that when it finally comes into motion, it won't just rise up, it will also rotate about its center of mass, as shown on the next page.

Fig. 1: about to move

Fig. 2: in motion

Notice how, on the right, a part of the rod is lower than where it was to begin with. For the most part, this kind of motion is not allowed. That's because these rods are resting on what is beneath them, which will be either more chain or the base of the pot. Either way, there's something preventing the left end of the chain actually descending. Instead, the rod simply pushes down on what's beneath it. We know from Sir Isaac Newton that every action has an equal and opposite reaction. So whatever is beneath the rod will push back up.

It's a bizarre conclusion but it seems as though the reason the chain rises above the pot is because it's pushed. By the pot!

Can we be sure this is the right explanation? No, we can't. But that's not something to be sad about. That's just the nature of scientific knowledge. A good scientist is never sure of anything. All we ever have is the best current theory.

So what makes a good theory in the first place?

Rule zero is that the theory should describe what we actually see in the world, which is almost too obvious to state. But the less obvious rule is that a theory should be testable. It should make predictions that we can go out and look for. We call this falsifiability.

One of the predictions of the Biggins and Warner description of the chain fountain is that the chain will rise higher above the pot if it has further to fall. Something we can test! I had the opportunity to do just that when I was invited on to the BBC's national prime-time TV chat show, *The One Show*, to talk about the beads experiment. As fortune would have it, there was a massive crane outside the studio on the day of recording, so we went up about 80 feet with 650 feet of chain in a pot to see what would happen. Normally, when I'm holding the chain fountain at regular standing height, it might reach about 6 inches above the pot, if I'm lucky. From the top of a crane outside the BBC's main London studio, Broadcasting House, we clocked 5 feet!

The highlight of this whole experience was watching Mark Warner talk about the chain fountain in an interview and hearing him refer to the phenomenon as the Mould Effect! I don't want to blow my own trumpet here, but Albert Einstein doesn't even have an effect and people say he's the best physicist ever.[H1]

I used to worry about having kids, because Mould is not an easy surname to have. Especially at school. Kids can be incredibly mean. But now, with the whole world talking about the Mould effect …

[H1] Not strictly true, there's the Einstein–de Haas effect, which reveals a relationship between magnetism, angular momentum, and spin of elementary particles.[S1]
[S1] That was mostly de Haas.

SMOKE GETS IN YOUR CDS

No party could be complete without some background music. But why stop at music? Why not get some background science going as well? For moments like these, you need to turn to your CD collection.

If you're too young to know what CDs are, they're a lot like MP3s, but MP3s that are round, shiny, and last for a maximum of 74 minutes. Advantages of CDs include their ability to be used as coasters. Disadvantages include having to build a lot of tiny shelves that are too small to be used for anything else.[H1]

Obviously, at a science party you're not going to use your CDs for playing music. What is this? The early 2000s? Play some music off the internet and turn those CDs into an experiment instead! Because it turns out that CDs are the perfect size and shape to make a toroidal vortex cannon, AKA a smoke-ring generator.

So raid your trash can and find:

1 An old toilet roll to form the main body of your vortex cannon.
2 Leaflets advertising your local pizzeria. You'll need a few stuck together to make them nice and stiff. If you're lucky, this process will have happened already at the bottom of your trash can.
3 Next, cut a hole the same size as the toilet roll in the middle of the pizza leaflet and slide it a quarter of an inch onto one end. Gaffer tape it in place, making sure you don't leave any gaps. This creates a structure that will hold the back window of your vortex cannon in place against one end of the toilet roll.
4 Tear some thin, flexible plastic from a food package. I like to use the top of a box that once contained fennel and butternut squash quiche with quinoa salad from our local vegan deli, but you may wish to use a different middle-class cliché to form the back window of your vortex cannon. Cut it to the same size as the pizza leaflets and gaffer tape it on, again don't leave any gaps.

[H1] Feel free to use this section of the book to explain CDs to any future generations, alongside your own explanations of where it all went wrong and why there are no polar bears left.

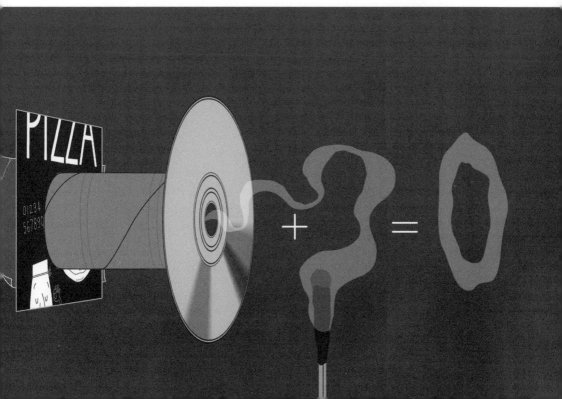

5 Finally, gaffer tape a CD to the other end of the toilet roll, being careful to get the hole in the exact center. For the last time, make sure you cover up any possible air gaps with yet more gaffer tape.

See previous page for an illustration of the whole process.

Ready ... aim ... smoke!

Once you've put your homemade vortex cannon together, you need to fill it with something you can see.

If you don't have a smoke machine handy—although that would surprise me, this is a science party after all—you have a couple of options for making your party rings. Get hold of some "smoke matches," the kind plumbers use to test gas pipes for leaks. They burn for about 20 seconds and produce heaps of smoke. To make the most of each one, hold the CD end of your vortex cannon with the hole pointing downward just above the burning match, to capture as much smoke as you can inside it before the match sputters out. Burning an incense stick will do the same thing but takes a lot longer and your party will smell like a gap-year student's bedroom. A snuffed-out candle also gives you a little stream of smoke to capture. A quicker alternative is finding someone who vapes.

Extra points are available if your party soundtrack matches the theme: two points for Smokey Robinson, five points for Deep Purple's "Smoke on the Water," ten points if you get Beyoncé to "Put A Ring On It."[H1]

Once you're fueled up with smoke, tapping gently on the plastic window generates a delicate ring-shape as the smoke emerges from the CD hole.

[H1] There are no points. Sorry.

You spin me right round, baby, right round[H1]

The inside of a toilet roll and the hole in a CD are just the right proportions to set a tiny doughnut of air spinning across the room—otherwise known as a toroidal vortex. It's not actually that hard to make a toroidal vortex. Pretty much any fluid[H2] that you send through a round hole will create a toroidal vortex, whether it's made by a homemade smoke-ring generator, a power-station chimney letting off steam, or a dolphin blowing air-doughnuts under water just for fun.

As it travels through the hole, the fast-moving smoky air gets a little "kick" from its interaction with the CD's inner edge. That —and the drag from slower-moving air just outside the hole— helps to get this plume of air turning back in on itself.

On and On and On and Smoke Cannon

What's great is that these doughnuts are incredibly stable. Once set up, the angular momentum of the spinning air, plus the lower pressure of the fast-moving air inside the ring versus the higher pressure of the air around it, means that they'll basically keep going until they hit something. Or eventually spin themselves out, thanks to some air-on-air friction action.

One of these teeny-tiny smoke rings will travel for about a yard. You can probably get it to go further if you practice, and also minimize air disturbances in the room. Like, if everyone at the party lies on the floor playing a postapocalyptic game of "Dead Polar Bears." That will really get the smoke-ring party started![S1]

[H1] Ooh, another one for the playlist. Thanks, Dead Or Alive!

[H2] Air is a fluid, from the point of view of a physicist or a toroidal vortex cannon.

[S1] These smoke rings are ... petite, to say the least. If you're a flash git like me, cut a circular hole in the base of a garden trash can and replace the lid with a sheet of garbage bag held in place with tape. Slap the garbage bag to push air out of the hole at the other end to make a giant vortex ring instead of teeny-tiny torus. You'll probably want to invest in a fog machine, otherwise you'll be powering through those incense sticks like there's no 9 a.m. lecture tomorrow.

SPINNING WASTEBASKET FIRE

WARNING

This experiment is dangerous. Only try this if you are experienced with fire and fire safety. We accept no liability. You do this at your own risk. Adults only.

Fire is dangerous. So are tornadoes. But do you know what's more dangerous than both of those things? Tornadoes made of fire. But it is possible to make one without injuring yourself or destroying your house. Why not try it out at your next garden party?

You will need

A metal-mesh wastebasket

A tealight

Lighter fluid

A barbecue lighter

A spinning plate, for example:
a lazy Susan or a record player

Fire extinguisher (CO_2
or powder)

STEP 1

Go outside. I want you to try this in your backyard, not your house.

STEP 2

Place your spinning plate on a flat, stable surface.

STEP 3

Remove the candle from your tealight, leaving just the metal container. This will hold your fuel.

STEP 4

Place a small amount of lighter fluid into the tealight case, no more than 1/16 of an inch deep.

STEP 5

Place the tealight case in the center of your wastebasket, then place the wastebasket in the center of your spinning plate.

STEP 6

It's important to makes sure everything is aligned before you ignite anything. So give the plate a slow spin and make any adjustments necessary to ensure the wastebasket and tealight case are centered.

STEP 7

Use your barbecue lighter to ignite the lighter fluid in the tealight case, then give the plate a gentle spin. After a few seconds, the small fire coming from the middle of the wastebasket should expand into a whirling tornado of fire! Nice!

What?!

Lighting a fire sets up strong currents of air around it. That's because as the air above the fire gets hot, it expands, becomes less dense, and rises. This is replaced with air drawn in from the sides. In our setup, the air coming in from the sides must pass through the mesh of the wastebasket.

When the wastebasket is set spinning, this air is given a little kick of angular momentum as it enters. It keeps that angular momentum as it heads toward the middle of the wastebasket, sucked in by the rising hot air. You can think of angular momentum as a measure of how much an object with mass is spinning around something. So the more massive the object is, the more angular momentum it has; the faster it's moving around the center of spin (the middle of the wastebasket in this case), the more angular momentum it has; and finally, the further it is from the center of spin, the more angular momentum it has. So to calculate angular momentum, you just multiply these three things together (mass, speed,[S1] and distance from center).

This is where a neat law of physics comes in: the number you just created, when you multiplied mass, speed, and distance together, can't change! You can change the constituent parts but when you multiply them together, you must get the same answer. This is called the conservation of angular momentum. What it means is, if the distance from the center goes down, for example, one of the other two values must go up to compensate.

That's what happens in our wastebasket. As the air moves toward the middle, the distance from the center goes down so the speed must go up to compensate. By the time it reaches the center its speed has greatly increased, resulting in a fast-spinning, fiery tornado.

[S1] It's technically the component of velocity perpendicular to the radius vector. I mean, angular momentum is a vector quantity anyway so I'm simplifying all over the place!

~~DOUBLE~~
SHE DOUBLE-BLINDED ME WITH SCIENCE

Science books are supposed to be educational.

So, concentrate!

If you need a little pharmacological help, try drinking a caffeinated beverage.

Time and again, research has shown that caffeine reduces reaction times and fatigue, increases concentration and ... err ... there's more but I've lost interest ... ooh, there's a pigeon on that roof over there! Hang on, I'll get a coffee.

Coffee break

Please tick your current state:

Caffeinated

Decaffeinated

Right I'm back! Where was I?

Yes. Caffeine: helping tired people function since 6 a.m. this morning.

But why should you take an extensive body of medical research's word for it? Why not use your next science-inspired social gathering to test it out?

It's cheaper than drugs.[H1]

So, it's time to open up those old experimental techniques you learned at school, and create a double-blind experiment to test the effects of caffeine on some willing human subjects.

[H1] Although caffeine is technically a drug.

IMPORTANT NOTE: Although many of us consume caffeine on a daily basis, it is still a stimulant and can affect people in significant and unexpected ways. Guidelines suggest that 400 milligrams of caffeine a day appears to be safe for most healthy adults—that's approximately four cups of brewed coffee, ten cans of cola, or two energy drinks. That's far more than you should consume in one go, if you choose to do this experiment.

ANOTHER IMPORTANT NOTE: Definitely don't do this experiment with the classic cocktail of vodka and energy drink. You're trying to create a double-blind test, not a double-blind-drunk one.

Now everyone's on the same page, let's get back to the lab report.

The point of doing a double-blind trial is to eliminate bias. By making sure that no one taking part knows who is receiving the active ingredient, you reduce the chances of influencing the result one way or another. That goes for all subjects involved and anyone administering the experiment. Randomized controlled trials are used to rigorously test new drugs and treatments, and are considered to be a gold-standard in evaluating the effectiveness of a medical treatment.

Coffee is the most obvious way to administer caffeine to your subjects. Other options include caffeine tablets, cola, and energy drinks, but none of those will lend a barista-style hipsterish quality to your experimenting.

To set up this trial, you need to ask your house-lab technician[H1] to lay out a series of cups marked with complicated codes and fill each of them with either caffeinated or decaffeinated coffee. The drinks should be as similar as possible in every other way, with no distinguishing features other than the codes. It's important that only the lab technician knows what is in each coded cup, and that they don't take part in or conduct any of the actual tests.

Now for the test: Your friendly lab technician will have randomly assigned one of the codes to each ~~participant~~ party guest already. Each party guest will need to consume their assigned drinks at the same time and at approximately the same speed. A traditional drinking song may come in useful at this point. Instead of "99 bottles of beer on the wall," try for a more ambitious number: "Infinitely many bottles of beer on the wall." The lyrics are the same for every verse, and it never ends. Your lab technician will tell you when to stop with the traditional closing words: "Please, please, for the love of all that is precious in the world, please no more singing."

Once the effects of the caffeine have kicked in,[H4] the classic way to compare reaction times is to find a long stick, hold it vertically, and drop it. It's best to have prearranged for one of your subjects to have their hand out to try and catch it as it falls, but if repeatedly dropping a stick onto the floor is entertaining enough on its own, you probably don't need caffeine or randomized controlled trials to have a good time.

For each of your subjects, measure how much of the stick falls through their hand before they catch it. The shorter the length of the stick that falls through before it's caught, the shorter the reaction time. You might want to repeat the test several times, and then do it again ten minutes later if the party mood

---

[H1] In the absence of a house-lab technician, pick the nerdiest of your friends and put them in charge of the preparation.[H2]
[H2] If you're looking around trying to work out which is the nerdiest of your friends, it's probably you.[H3]
[H3] If you've already prepared a spreadsheet logging the nerdiest characteristics of your entire social group, it's definitely you.
[H4] Or not, depending on which group you were randomly assigned to ...

is flagging because all your CDs have been turned into vortex cannons and your wastebasket is on fire in the backyard.[H1]

Log the average catch length for each person on a table, and use the conversion chart below to turn those lengths into reaction times. Then ask your lab technician to reveal the caffeine content of each party guest's coded cup. If you're feeling really science-y, display your results on a chart, with caffeine level along the bottom and reaction time up the side.

Length-to-reaction time conversion chart

Catch distance (inches)	Reaction time (milliseconds)	Catch distance (inches)	Reaction time (milliseconds)
1	72	13	260
2	102	14	269
3	125	15	279
4	144	16	288
5	161	17	297
6	176	18	306
7	191	19	314
8	204	20	322
9	216	21	330
10	228	22	338
11	239	23	345
12	249	24	353

If you've science-d successfully, your results might show that caffeine shortens reaction time. They might not. It may be difficult to tell for sure. If in doubt, blame the lab technician.

Luckily, there's plenty of research out there to show what happens when humans get caffeinated. It's thought that the graph of stimulation level versus performance at any given task follows a hump-shaped curve: poor performance can occur when the subject is undercaffeinated, but also when they are

[H1] See pages 146 and 150.

overcaffeinated. The best place to hang out is somewhere in the middle.

Unfortunately, if you're a habitual coffee drinker it can take twice as much caffeine to see any significant effect. And as high intake has been linked to irregular heartbeats, nausea, and insomnia, you might not want to push the envelope too much with this home experiment. Instead, why not increase the effect by giving up caffeine for a week or two before doing this experiment? Oh yes, I know why ...

COFFFFFEEEEEEE

Is there another way?

It looks like caffeine may not be the only way to improve concentration. The long queue for the toilet at your house party might have a similar effect, as shown in a 2011 research paper from Tuk, Trampe, and Warlop: "Inhibition Spill-Over: Sensations of Peeing Urgency Lead to Increased Impulse Control in Unrelated Domains." It turns out that a little bit of needing to go to the toilet might actually help you focus on a given task.

A 2007 BBC documentary claimed that Conservative Party leader David Cameron used this technique during an hour-long speech at that year's Tory Party conference. The speech, delivered after he "specifically refrained from using the toilet," was seen as a triumph, and three years later he was elected Prime Minister. Hopefully he had a chance to go before that happened.

The 2011 study backs up these claims, but it also found limits to the method's effectiveness. Like the caffeine curve, if you get beyond a "point of no return," your concentration worsens. With the diuretic properties of all the coffee you've been drinking, you might find yourself at the upper end of both those curves sooner than you expect. Now excuse me while I pop to the bathroom ...

SCIENCE COCKTAILS

Get into the party spirit—and by spirit, yes, we do mean alcohol —with a bevy of scientific beverages.

Responsible adults only!

Irresponsible adults can sit in the corner with a sippy cup of weak juice and watch the responsible adults make idiots of themselves instead.

Gin and (Light) Tonic

A stiff gin and tonic is the preferred drink of many off-duty scientists. Not just for the complex range of botanicals in the gin, but also for its bizarre blue glow under ultraviolet light.

You might have spotted it happening accidentally ... while drinking a classy G&T in a dirty dive of a nightclub with UV lights, or that time you were caught drinking "mother's ruin" in the grocery store checkout queue one Friday lunchtime, and it caught the glow of those little blue lights they use to detect banknote fraud.

It's not actually the gin that glows, but the tonic. The quinine that lends its bitter taste to the mixer is fluorescent. Ultraviolet light, with its tiny wavelength beyond the blue end of the visible spectrum, is undetectable by the human visual system. That's one thing bees have got that we don't, but they do find it more difficult to handle a tumbler with their little legs.

Back to the tonic. When that "invisible" UV light hits the quinine molecules, it's absorbed then reemitted. But the light it spits back out is a longer wavelength than the original UV—so we see it as a visible blue glow coming from inside the glass.

Quinine, or the cinchona bark it comes from, has been used since the 17th century for treating and preventing malaria. If you're thinking a couple of cheeky G&Ts might keep the disease at bay on your next vacation, be warned that the levels in standard tonic waters are so low that you'd be drinking about 6½ gallons of the stuff every day. That's one heck of a preventative measure.

9fl oz gin and tonic
For medicinal use only
Drink 100 daily
Side effects may occur

Go with the Flow

How about a bit of convection with your cocktail? For this you'll need a nice thick liqueur such as Kahlúa and some light cream.

Your drinking receptacle for this tipple will be a dinner plate.

Pour the liqueur onto the plate to the depth of a quarter inch, then carefully spoon the cream on top. If you have a pipette handy, even better.

You should start to see tiny cells forming: regions of liqueur surrounded by loops of cream.

Look closely at those loops of cream and you'll notice they are spinning.

What you're seeing is convection, but not the kind of convection you're used to.

If you put a pan of water on the stove, the water at the bottom will heat up, expand, become less dense,

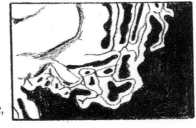

and rise to the surface. There, it will cool down again in contact with the air, contract, become more dense, and fall back down to the bottom where the cycle continues.

This cocktail convection is different. It's an example of solutal convection. Alcohol evaporates from the Kahlúa at the surface, making it more dense and causing it to fall to the bottom. This is replaced with more alcoholic Kahlúa from below, which evaporates and falls, and so the cycle continues. It's a little more complicated than that, and involves the cream and surface tension, but that is "beyond the scope of this book," as the professionals say. Just drink the damn cocktail, OK.

Strawberry DNA-quiri

Behold! The key to life! The holder of our heredity! It's DNA, but it's drinkable!

You might have seen this at a science fair, or watched a video online that shows you how to extract DNA from your cheek cell using dishwashing liquid, salt, enzymes, and a layer of alcohol. Mmm ... that cocktail sounds ... delicious?

Don't worry, there is an edible version! A bunch of brilliant San Francisco biohackers have come up with a recipe that actually extracts strings of DNA for you to see with your naked eye *and* tastes delicious in a martini glass.

I've adapted it slightly here. It's a little complicated, but not much more than the average cocktail, and the results are far more scientifically satisfying.

First, get yourself a plastic ziplock bag and pop a handful of frozen strawberries and some pineapple juice inside. Gently crush them up together with your hands until you have a nice smooth consistency. The freezing is important—it helps to break down the cell walls inside the strawberry. That allows the enzymes in the pineapple juice to sneak inside the fruit cells and raid the nucleus for DNA. It's important to gently crush it, rather than use a powerful electronic blender. You want to break down the cell walls, a process technically known as "lysis," but not mangle the DNA inside.

You can try other fruits in this cocktail recipe, if you prefer. Kiwi and banana are good suggestions, as they are both polyploid: that is they have more than the usual two sets of chromosomes in each cell. Bananas have three, and the kiwi comes with six as standard. But when you're only using edible cocktail ingredients to extract DNA, you need as much as possible in the first place. Strawberries are the connoisseur's choice as they are octoploid, with eight copies of the genome in every cell. Yum.

Now pop your bag of crushed fruit into a bowl of water heated to around 122°F (50°C). No thermometer? Don't worry too much, just get it hotter than bathwater, but not so hot it scalds your finger. Heat the ziplock bag of fruit in the hot-water bath for ten minutes, then place it in an ice bath for ten minutes. The gentle heat helps release even more DNA from the broken cells, but it's easy to let it go too far and start to break down the DNA itself. The ice bath works to slow down the process so you still have something to see at the next stage.

Finally, strain your pulpy mess through a strainer into the bottom of a glass. Carefully layer some strong, ice-cold alcohol on top by pouring it in via the back of a spoon. Overproof rum works well, or anything else that's reasonably clear with a high alcohol content. Because the lengths of strawberry DNA dissolve better in the watery fruit mush than in alcohol, they begin to "precipitate" before your very eyes. Almost immediately you'll see strands of DNA creeping out of the fruit layer and into the clear alcohol layer.

To complete the world's most science-y cocktail, add a tiny umbrella and use it to twirl a few strands of DNA up and out of the glass. It looks a lot like boogers but don't let that put you off. When you're done marveling at the very core of our existence, add sugar syrup to taste, swirl your glass around to mix it up, and swallow in one. Ain't the secret of life itself grand?

pH is for pHave another pHint of pHort

If you've been wanting to test the pH of various household products since reading about my date night in Chapter 2,[H1] but don't have any noodles to hand, don't worry. The drinkable version is here to save you.

Unlike red cabbage juice, which is a great pH indicator but a terrible cocktail ingredient, there are some other tasty beverages out there that do the same thing. Port wine, and a very particular British beverage called black currant squash, a sugary cordial which you normally dilute with water to make a delicious fruity drink. You might be able to get hold of some from a farmers' market or specialist British grocery store. Both of these drinks contain the same group of chemicals as red cabbage: anthocyanins. That's what turns your drink purply-blue when exposed to an alkali such as baking soda, and pinky-red with an acid such as lemon juice.

If you fancy combining two experiments in one, you can mix the gentle UV glow of tonic water with a pH-indicating gin. Look out for a select few brands of gin that contain pea flower, another natural source of anthocyanins. Light blue in the bottle, it turns pink when you add a slightly acidic tonic water mixer. Magic![S1]

[H1] See page 56
[S1] Not magic. Science.

Lazy Lava Lamp

It's not always good to drink and science, so for our final cocktail we've got something nonalcoholic. It's also completely undrinkable. But it does look properly science-y.

Re-create that smooth Seventies cocktail-lounge vibe by layering up liquids to make a homemade lava lamp. Groovy, baby!

This "highly experimental cocktail"—AKA, not safe for human consumption—works best if you use two liquids that have very different densities. Fill half of a pint glass with high-density, sugary black currant squash, then top it off with about 3 inches of much lower density vegetable oil, to create two different blocks of color at the top and bottom.

Once you've layered up, drop a crushed soluble vitamin C tablet in, to start the lava-lamp effect. As the chunks of vitamin tablet sink, they react with water in the squash to produce carbon dioxide gas. Those tiny gas bubbles form around the tablet chunks at the bottom of the glass, gradually joining together into bigger bubbles until they rise up through the squash. As they carry on rising into the oil layer, they drag tiny chunks of tablet and little rivers of red liquid with them. At the surface, those bubbles pop and the chunks drop back down, taking some of the oil layer with them, to create a writhing, bubbling volcano in a glass.

And if you used a strong black currant squash, full of all those lovely anthocyanins from earlier, you might see it change color. Some of the carbon dioxide is absorbed into the water, turning it more acidic and dropping the pH. Bonus science!

You still can't drink it, though. Not even with all that vitamin C.

UNIVERSE STUFF

6   Chapter

6

166/189

Like most nerds, we're fascinated by the solar system we inhabit and everything outside it. From stargazing to gravitational waves, we've captured some of our favorite corners of the Universe[H1] and put them on these pages so you can become a space traveler from the comfort of your armchair.

We'll be looking at space from Earth and Earth from space. And we'll attempt to predict the fate of everything that exists out there, by staring down the arrow of time that is hidden in the very laws of thermodynamics.

But first, let's take a rocket into low Earth orbit. And, from this vantage point, explore a new world of mathematics.

Onto the next page in T minus 5 ... 4 ... 3 ... 2 ... 1 ...

[H1] Quite a challenge, as the ever-expanding curved surface of the universe has a distinct lack of corners.

HOW THE COASTLINE OF BRITAIN GAVE BIRTH TO NEW MATHEMATICS

How long is the coastline of Britain? You don't know, of course, but you can Google it. The problem is trying to find an answer that everybody agrees on. The Ordnance Survey says it's 11,073 miles. But the CIA (who also take an interest in this sort of thing, apparently) says it's 7,723 miles. So who's right?

Let's do a bit of our own measuring and see how we get on. Here's a segment of the coastline of Scotland. For your convenience, I've laid down a load of giant rulers, 10 miles in length.

All you have to do is count the number of rulers, which in this case is 11, giving us a length of 110 miles. You might be concerned that these giant rulers are too big; that they don't get into all the little nooks and crannies. So let's try again with a smaller ruler. Let's go for one half the size this time: 5 miles.

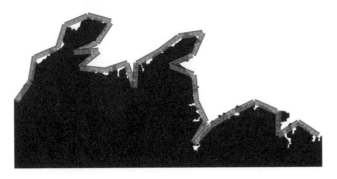

We now have 30 rulers, or 150 miles. That's an extra 40 miles compared to our previous measurement! And it turns out, the smaller the ruler you use, the bigger the answer you get. So the question becomes, how small does your ruler need to be to get the right answer? How small does it need to be to get into all those nooks and crannies? And here's the problem, there are nooks and crannies at all scales down to the atomic level.

You can imagine making your ruler smaller and smaller but always finding smaller and smaller nooks and crannies that your ruler is skipping. Until eventually your ruler is the size of a molecule, and then how would you draw on the numbers?

A mathematician called Benoit B Mandelbrot[1] was interested in this coastline measuring problem and decided to take a mathematical approach.

How might you construct a perfect mathematical coastline? Let's start with a straight line.

This doesn't match natural coastlines very well because natural coastlines have nooks and crannies; they have knobbly bits. So let's add a knobbly bit to our coastline.

This is looking better. But we're still not there. If we zoom in on our knobbly coastline, we find short sections of straight coastline. But in the real world, there's knobblyness at all levels, so let's add knobbly bits to these straight lines.

This has just created a load of smaller straight lines. So let's keep adding knobbly bits to straight lines. What happens if we do that forever? We get  something like this:

This is called a Koch snowflake, but it also does a good job of being a coastline, if a little too symmetrically.

You might have noticed that this coastline is "self-similar."

[1] The B stands for Benoit B[2] Mandelbrot.
[2] The B stands for Benoit B[1] Mandelbrot.

In the same way that a floret of broccoli looks like a whole broccoli in miniature, you can see how segments of the Koch snowflake look like the whole thing in miniature. We can use this self-similarity to our advantage.

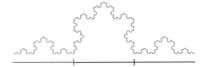

Notice how the segment in red in the figure above looks like a baby version of the whole thing. If you take a ruler to it, you'll find that the whole thing is three times wider, and therefore three times longer than the segment in red.

But there's another way to look at it:

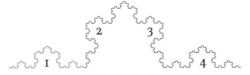

See how the whole thing is made up of four identical segments like the segment in red. So it seems as though the whole thing is actually four times longer than the segment in red!

Depending on how we approach the problem, the whole coastline is three times longer than the segment in red *and* four times longer!

In other words: 3 = 4.

When you get an answer like that, it usually means you've made a mistake. But very occasionally it means you've invented a new branch of mathematics, so let's see if we can find our way out of this conundrum.

For a start, we're not really saying 3 = 4, we're saying that when you multiply the length of the red segment by 3 you get the same answer as if you multiplied it by 4.

There's only one length of coastline for which that is true: 0 miles (3 × 0 miles = 0 miles and 4 × 0 miles = 0 miles). So we can finally, conclusively say that our mathematical coastline—and, by extension, the coastline of Britain—is 0 miles long. Except that I've been to the coast and it's definitely longer than that.

There is another solution, but it's not a proper length in the traditional sense. It's not even a number. I'm talking, of course, about infinity. Multiplying infinity by anything just gives you infinity. So it meets our criteria but it's a bizarre conclusion, to say the least. With a little more rigor, this is actually how you prove the Koch snowflake is infinitely long. It's not a perfect analogy for the British coastline because the knobblyness does stop eventually when you get down to atoms. But what it does tell us is that trying to measure the coastline of a country is a pointless task. There is no meaningful right answer.

Mandelbrot formalized the study of these self-similar geometries into a new branch of mathematics called fractals that gives rise to wonderful images like the one below:

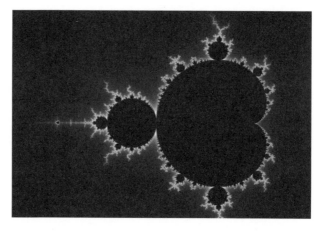

It also gives us the language we need to describe such geometries. So while we can't say how long the coastline of Britain is, we *can* describe its knobblyness mathematically. It turns out these curious objects have, in some sense, noninteger dimensions. Where a line is 1 dimensional and a sheet is 2 dimensional, the Koch snowflake is 1.26 dimensional. It's somewhere between a line and a sheet. How bizarre! And, just so you know, the coastline of Britain is about 1.2 dimensional.

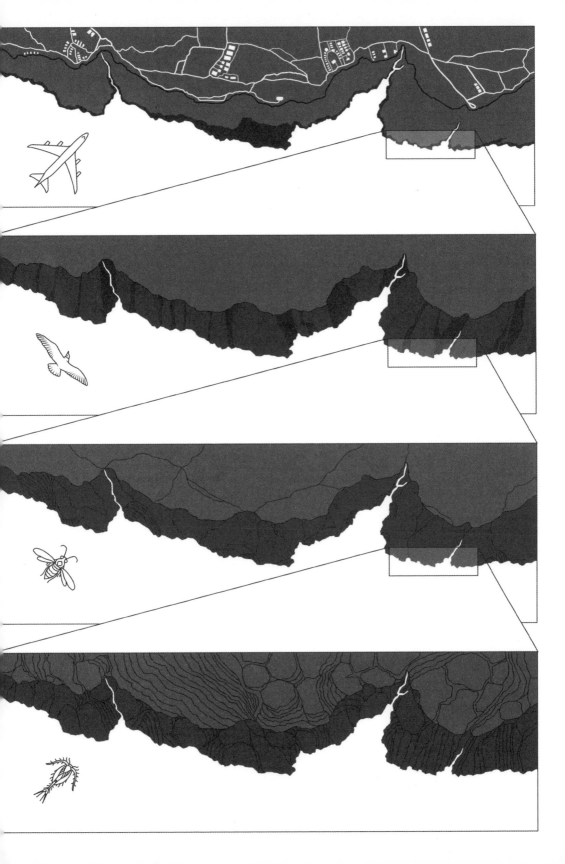

A LATE-NIGHT THOUGHT EXPERIMENT

Have you ever dreamed of flying through space?

Well, you are. Right now.

As you lie there on your flatpack Ikea bed[H1] reading this book all curled up and ready to sleep, the Earth is rotating on its 23 hour, 56 minute, 4.1 second axis. If you're lucky enough to be in a bed at the equator, that means you're moving at around 1,000 miles an hour. The speed drops as you get close to one of the poles, as if you were in exotic Madagascar. Or ancient Peru. Or ... Scotland. Whichever you prefer.

Your speed is all relative to a fixed point, of course, with that fixed point being the centre of the Earth.

Then again, as you snuggle down under your covers to avoid the existential crisis bubbling away inside your skull, is the center of the Earth really a fixed point?

The Earth is in an almost perfectly circular orbit around the Sun, so that "fixed" point is moving at around 67,000 miles an hour, give or take a few thousand.

Now take your wide-eyed gaze out of the window. You might as well, as there's no way you'll be sleeping tonight. Allow your eyes to adjust to the soul-searching darkness of the night sky. If there's no moon, and you're away from the light pollution of a city, around 15 or 20 minutes should be long enough to reach a state of total

[H1] According to every journalist who has ever tried to write about the furniture industry and make it interesting, one in ten Europeans are conceived in an Ikea bed. Some articles suggest it's one in five. That's double the "sexing up." Either way, approximately 884 million people a year visit Ikea in 50 countries worldwide. It's not known how many of those are sneaking into the showroom at night to test this mysterious hypothesis.

metaphysical ennui. I mean, long enough to make out the Milky Way, a glowing band of stars arching across the night and forming part of our own galaxy. Which, in its own way, is turning.

Lie rigid with your head digging into a soft pillow, paralyzed by the insignificance of our tiny Earth's place in this vast galaxy, with our part of its spiral arms swirling around a massive central disk of stars at a mind-numbingly fast 515,000 miles an hour, or 143 miles a second. That's twice as fast as lightning, more or less.

In a few hours, days or possibly weeks, the transcendental melancholy you feel at every bedtime will fade.

Just in time to realize that the Milky Way itself is moving, and has been since the moment of the Big Bang. Contemplate the incomparable unlikeliness of your existence here in a bed, on a rock, spinning around and flying through space as our little patch of Universe expands away from the dawn of time itself at a comfortable 1.3 million miles an hour. And think back to those innocent times of the previous page, when you merely dreamed of flying through space.

Relatively, you're already a space traveler in your sleep.

Goodnight.

WE NEED TO TALK ABOUT ENTROPY

You may have heard entropy described as a measure of disorder. The more disordered something is, the more entropy it has. And you may have heard that entropy is always increasing, so that the disorder in the Universe is always going up.

It's a popular definition because it's easy to understand and it can be seen somewhat intuitively in everyday life—it's hard to keep a house tidy and now you know why, you're fighting the general tendency of the Universe to become disordered.

I'm not a big fan of this definition though, because I don't think it helps us to understand what entropy is. So I'm going to share a far superior definition with you that goes back to why the notion of entropy was invented in the first place and what it might tell us about the fate of our Universe.

During the industrial revolution people became obsessed with the efficiency of engines. A mathematical language was developed to probe the challenge in a systematic way, and it was in this environment that the concept of entropy was born.

A simple engine is something you put fuel into, where it is burned to produce heat. That heat is then somehow turned into motion. This type of engine is called an internal combustion engine but it's not the only type, and the engine I'll use to illustrate the concept of entropy is actually the other type—an *external* combustion engine called a Stirling engine.

The Stirling engine cycle

10 When the air inside the chamber is in contact with the hot bottom plate,[S1] it heats up and expands. This pushes on the piston which turns the flywheel.

20 The flywheel is also connected to the displacer inside the chamber, so that when it turns, the displacer moves down.

30 The air in the chamber is now in contact with the cold top plate.

40 The air in the chamber now cools down and contracts. This pulls on the piston allowing the flywheel to keep turning.

50 The turning flywheel pulls the displacer back up so that the air in the chamber is now in contact with the hot plate once again.

60 GO TO 10.

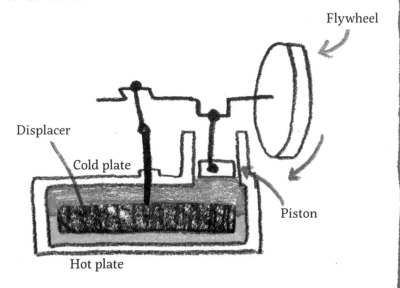

Flywheel

Displacer

Cold plate

Piston

Hot plate

[S1] Hot bottom plates, you make this rockin' wheel go round.

The Stirling engine and entrophy

The key to the operation of a Stirling engine is not just to heat the thing up but to have a *difference* in temperature between the two plates. You can even run a Stirling engine on ice pressed against one of the plates, so long as the other plate is at room temperature (or indeed any temperature, so long as it's different). The reason it's called an external combustion engine is because you could burn something outside the engine to heat one of the plates.

How does this relate to entropy? Imagine that you have two slabs of metal, one hot and one cold. When you press the two together, you know from experience that heat will flow from the hot slab to the cold slab until they are both the same temperature. At this point nothing else will happen. Without outside intervention, this is the end of the game. We never see that process in reverse—if two objects are at the same temperature, we won't see one object spontaneously steal heat from the other.

Let's now imagine we use these slabs to power our Stirling engine. We put the cold slab against the top plate and the hot slab against the bottom plate.

In the beginning, the heat energy is all clumped together in the hot slab. Heat then flows from the hot slab to the cold slab *through* the engine causing it to turn! But once the heat energy is evenly distributed between the slabs; once they are the same temperature, the engine stops.

The take-home lesson is this:

- Energy is only useful when it's clumped together.
- When you use that energy to do something useful, it spreads out.
- Once it's spread out, you can't use it anymore.

So the engineers and mathematicians of the industrial revolution came up with this concept:

ENTROPY: A measure of how spread out your energy is.

So to say that entropy always increases is to say that energy is always spreading out.[S1]

Our inevitable demise

Extrapolating this idea, we end up with something of a doomsday scenario. We will eventually reach a time when all the energy is spread out and our engines will stop running, including the engines of our bodies.

It's not all bad news. Fortunately for us earthlings, there are loads of sources of clumped-together energy that we haven't used yet. Things like coal, oil, and gas. When we burn those fuels we're running our engines and spreading that energy out. And once we do, that's it, we can't use it anymore. They are nonrenewables. Luckily there is still one giant source of clumped-together energy at our disposal: the Sun. Once the fossil fuels have run out we can continue to power our engines using solar panels. Not just solar panels, though. Hydroelectric power, wind power, and biofuels are all, ultimately, derived from the power of the Sun.[S2]

I'm not saying we should use up our fossil fuels before switching to the Sun. Quite the opposite, we should switch to Sun-based energy clumps as soon as possible. I hardly need to say that burning fossil fuels has one or two unfortunate side effects.

[S1] It is possible to clump energy together, but only at the expense of energy spreading out elsewhere. Our little Stirling engine is a good example of this because it's reversible: if you manually turn the wheel it will "pump"' heat from one plate into the other, clumping the energy together on one side. But to turn the wheel you must use your muscles, which generate heat energy that spreads out from your body.

[S2] Hydroelectric power harnesses the kinetic energy of water flowing downhill. But the water only got up the hill in the first place by evaporation from the ocean, a process powered by the Sun. Winds form when air flows from areas of high pressure to areas of low pressure, again powered by the Sun. And biofuels are derived from plants that lock up the energy of the Sun in chemical bonds through photosynthesis.

But even the energy in the Sun will spread out eventually, as will all the energy in the Universe. And once all the energy in the Universe is evenly distributed, nothing interesting can ever happen again. This is known as the heat death of the Universe and it's our best guess at how the Universe will end. It won't happen for another ten thousand trillion trillion trillion trillion trillion trillion trillion trillion years, though, so you can relax.

Why *does* entropy always increase?

The short answer is that entropy is more likely to increase so that's what we see! But why is it more likely?

Consider a box with a number of ball pit balls inside, the kind you get in soft-play centers. Just enough for one layer of balls. There are two colors, black and red, and they have been arranged so that all the red balls are on one side and all the black balls are on the other.

Now imagine closing the lid and giving the box a good shake. When you open the box again to take a look, how do you expect the balls to be arranged? You wouldn't be surprised to find the two colors were now mixed up. You would be surprised to find them arranged just as they were at the start.

That's because you have an intuitive understanding that entropy increases; that a jumble of balls doesn't spontaneously arrange itself into segregated groups. Your intuition no doubt comes from experience, but there's a statistical explanation behind it too. There are very many ways to arrange a load of ball pit balls in a box that would be considered jumbled. And only a few ways that would be considered orderly. So we *expect* to find a jumbled state because it is more likely. This is where the notion of entropy as a measure of disorder comes from. It takes effort to arrange the balls in an orderly state and it takes effort to keep a house tidy. But really, this box of balls is an analogy for what's going on at the molecular scale. It's why farts diffuse through rooms and why stirring milk into tea mixes the liquids together. And it's why you'll never see the reverse: milk never unstirs and farts never regroup around your butthole.

There is a very slim chance that the ball pit balls could find themselves in an orderly state after a jiggle, but the likelihood goes down as the number of balls goes up. The same is true for atoms and molecules, and when you're dealing with something the size of a fart, the chance of spontaneous order arising becomes astronomically low.

So far as we know, the physical laws that govern the Universe work equally well in reverse. If you were able to film the interactions of subatomic particles, for example, then play the movie back to a friend, there would be no way for your friend to tell if you were playing the movie forward or in reverse. It's only when you zoom out and look at many particles interacting that you can discern a direction of time. Only when there are enough interactions for you to consider them statistically, can you discern a clear future and past. In fact, all the phenomena that we associate with the passage of time can be traced back to entropy and its inexorable march upward. So in a sense, time itself is nothing more than applied statistics.

INTERSTELLAR TRIP ADVISOR

Time is indeed marching on, and it's worth having a think about what the options are when we can no longer stay on the only Earth we've ever known,[H1] home to the only intelligent life in our Universe.[H2]

Here's a handy guide to three of humanity's potential new homes.

Trappist-1e

#1 IN EARTH-LIKE DESTINATIONS (SINCE 2016)

RATING: ★ ★ ★ ★ ★

- - - - - - - - - - - - -
LOCATION: Constellation of Aquarius
- - - - - - - - - - - - -

Suggested journey time from Earth: 39 years at light speed

LOCAL ATTRACTIONS

1. Soothing red light from a dwarf star 2,000 times dimmer than the Sun.
2. "Tidally locked"—one side always faces the sun, for topping off your tan at any time of day or night.
3. Six other Earth-size planets in the same solar system, for the ultimate multitrip destination.

[H1] Probably.
[H2] Possibly.

TOP REASON FOR VISITING

With a slightly smaller radius than Earth, it's likely that the effect of gravity will be slightly less, giving you that light and carefree vacation feeling as soon as you arrive.

REVIEWS

"Great little find from 2016"

I wasn't aware of Trappist-1e until earlier this year, and already it's my vacation planet of choice. The whole Trappist solar system would fit well within Mercury's orbit around our own Sun, so it's easy to cut down on local transport costs if you're on a budget.

[ColoradoWolf]

"I spent a year on Trappist-1e and I wish I'd stayed longer"

Due to its small orbit, one year on the planet turned out to be only six Earth days. Not long enough for this great little rock. Despite one side facing the sun and the other left in shadow, there's a sweet little line in between which is just the right temperature for liquid water. Forever lingering at that perfect moment between day and night, it's always cocktail hour on Trappist-1e!

[lgm83]

Book your flight, hotel, and cryostasis pod together to save $$$s

Kepler-452b

#2 IN EARTH-LIKE DESTINATIONS (PREVIOUSLY #1)

RATING: ★ ★ ★ ★ ☆

LOCATION: Constellation of Cygnus

Suggested journey time from Earth: 1,400 years at light speed

LOCAL ATTRACTIONS

1. A geologist's dream, this "supersize Earth" might offer active volcanoes to help your trip go with a bang.
2. With an equilibrium temperature close to Earth's, liquid water is highly likely (don't forget your swimsuit!).
3. 385-day orbit gives a familiar length of year, for that real home-from-home feeling.

TOP REASON FOR VISITING

Winner of the 2015 "most Earthlike planet" award.

REVIEWS

★ ★ ★ ★ ☆

"Worth the 26-million-year journey to visit Earth 2.0"

Lives up to its name as a "Super Earth"—it's 60 percent bigger and we think 60 percent better. A great destination for all the family, or at least those that survive the trip.

`[nasa _ TESS]`

★ ★ ★ ★ ☆

"Fabulous getaway for years to come"

There may be exoplanets closer to Earth with more liquid water but there's something special about Kepler-452b. Possibly its larger mass, which will combat runaway greenhouse effects and keep this cool little planet habitable an extra 500 million years longer than Earth. We'll be back for sure!

[JamesWebb]

We check for the lowest prices so you don't have to!

Economy flights (one way, scientific instruments only) from just $10,000,000,000

PSO J318.5-22

#4,298 IN EARTHLIKE DESTINATIONS

RATING: ★ ★ ★ ⯨ ☆

LOCATION: The constellation of Capricorn

Suggested journey time from Earth: 75 years at light speed

LOCAL ATTRACTIONS

1. A "rogue planet" with no host star, the nightlife here carries on 24 hours a day.
2. Clouds of 1,500°F dust and molten iron create a unique atmosphere.
3. Only 20 million years old, making it the hottest new destination for the adventurous traveler.

TOP REASON FOR VISITING

This unique world is the size of Jupiter so get ready to explore! Since its discovery in 2013, it's become the perfect place to get away from it all. And when we say "all" we really mean it …

REVIEWS

"Unique"

There are lots of free-floating objects out there in space, but none of them quite as much like Earth as PSO J318.5-22. That is to say, it's nothing like Earth. At all. Still, we've got some great memories we'll be carrying with us at least 20 million years into the future.

P.S. Don't skip leg day before you go. The place is about 2,500 times the mass of Earth. Makes it kinda heavy on the gravity front.

[telescoped49]

"Don't pack light if you want to enjoy yourselves"

My husband and I took a minibreak on PSO J318.5-22. It wasn't totally clear in the brochure that you need to bring your own hi-tech life-support systems. The basic scuba gear we had with us burst into flames on contact with the planet's atmosphere. Hubby's heat rash was quite the sight! Put a bit of a damper on the trip but I suppose it's something else to talk about when we get back home, if any of our friends are still alive that is.

[exo _ xxx]

> Dress to impress, and to avoid deadly radiation in transit! Lead-lined swimsuits now available to hire.

SPACE IN YOUR FACE

Here are some tips on how to see beyond our Earth while staying comfortably indoors.

Tune in to the early Universe

It's time to re-create some Nobel prize-winning physics in your own home! First, find yourself an analog TV. If there isn't one in your house, try the 1990s.

Now, switch it on, grab some popcorn, and watch the black-and-white fuzz of static dancing on the screen. Detuning an FM radio[H1] to find the hissing noise between stations will get the same effect. That's evidence of the Big Bang right there. Around one percent of the interference you're seeing or hearing is leftover traces of radiation from the birth of the Universe.

Analog TVs and radios are both sensitive to electromagnetic "noise"—unwanted transmissions from Earth and from space picked up by their antennae—and also noise from inside their own electronics. When you're not tuned to a strong signal from a particular station, all that other junk shows up instead. And a tiny bit of that junk is proof of the Big Bang.

Cosmic Microwave Background, or CMB, is like the proud first photograph of a baby. A baby Universe, though. Aww! Since it popped into existence, that Universe has been cooling and expanding into the vastness of space. CMB is the remnants of light from the moment hydrogen atoms were first created in an approximately 5,000°F soup of matter and energy, just 400,000 years after it all kicked off. And just like the rest of the Universe, that light has also been expanding and cooling for the last 13.8 billion years, dropping from about 5,000°F to just 5°F above

---

[H1] Like most things today, there's an app for that. Your smartphone most likely has an FM radio chip installed, though it might not be activated. Worth taking a look as it might come in handy during an apocalypse.

absolute zero, the coldest possible temperature that anything can be. And instead of staying as visible light, it stretched as it cooled to become microwaves, which is why your retro receiver kit can detect it.

By tuning their increasingly powerful radio telescopes and probes to look across the entire sky, astrophysicists can map out the peaks, troughs, and patterns of this radiation. The results are an almost perfectly smooth spread of microwaves in every direction, which is what you'd expect from a uniformly expanding Universe. But by looking really closely, scientists have detected tiny, microdegree differences in its temperature all over the place. The data is full of clumpy bits and sparse bits, showing that there were tiny variations in the density of the Universe in the first moments after the Big Bang. Slightly denser parts of the early Universe tended to stick together and attract even more stuff to join in the fun, which in turn became stars, galaxies, and all the matter that we know. If you've ever wondered why scientists can be so sure about how the Universe began, how old it is and how it is likely to end, this CMB data is a big part of what makes them so confident.

This lingering echo of the Big Bang was first identified by Arno Penzias and Robert Wilson, after they spent years painstakingly scanning the sky with the latest in radio astronomy equipment, the Holmdel Horn Antenna in New Jersey. At first they thought the background "hum" in their data was bird poop[H1] from the pigeons nesting inside their equipment, but after removing the birds (and their poop) the noise was still there. After all other possibilities had been ruled out, they had to accept that they were detecting CMB, a signal from the dawn of time, which pretty much proved the Big Bang Theory and won them the Nobel Prize for Physics in 1978.

If only they'd taken an evening off, and switched on the telly instead. It would have saved them a lot of time and rubber gloves ...

Even if you're not tuned in—or rather tuned out—just hold your hand in front of your face instead. Your fingertip has about 400 of those original photons from the Big Bang racing through it at any one time.

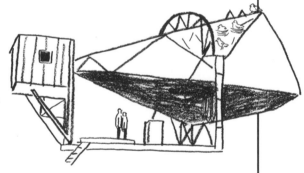

Since the UK went from analogue to digital TV it's become a lot easier to peer into the first moments of the Universe, if you have the right equipment— because you don't get any of that nasty interference from actual television programs. *The Big Bang Theory* might not be available on analog anymore, but the Big Bang still is. That's real scientific progress for you.

[H1] "White dielectric material," if you want the science-y words for pigeon poop.

187

SPECIAL GUEST FEATURE

And now, a short contribution from another celebrity science performer. They've just completed a world tour or, more accurately, the world has just completed a tour of them. They're an absolutely stellar talent, eclipsing all others in their field, so please welcome ... the Sun!

Thanks, Nerds.

It's great to be contributing one of my poems to this book, especially as it's written by a couple of Brits. They're not too familiar with my work.

There's a reason I took this unpaid opportunity to write a little something for Helen and Steve. It's becoming more and more obvious that all of you on Earth don't appreciate me enough.

You used to worship me, back in the good old days. But now you're all too busy searching for Earthlike planets orbiting other solar systems. Too busy watching distant galaxies being pointed at by floppy-haired Mancunian physics professors. You know who I mean ...

If I'm being perfectly honest, I've had enough of being ignored. To put it another way, This Sun Has Got Its Huff On.

---

* In fact two British tabloids: The *Sun* and also The *Star*.
** Actual genuine name of a star.********
*** Which is all of One Direction.
**** I'm not totally sure I've pronounced that correctly, but hey, I'm the Sun, what do I know?
***** Or 865,000 miles if you want to be all imperial about it.
****** And become 200 times my current size, then you'll really notice me! After I've swallowed the Earth I'll likely shrink back down into a white dwarf. But don't worry, that's about four or five billion years away from now. You've got other stuff to worry about in the meantime.
******* Seriously, I've only got nine likes ... Well, it's gone down to eight now since Pluto defriended me.
******** Wow, these asterisks look like stars! Awesome!

The Ballad of the Lonely Sun

I used to be someone
Now I'm just another sun
One of a hundred thousand
Billion billion

You treat me insignificantly
Name a tabloid after me[*]
Synonymous with paparazzi
Just a backdrop for Brian Cox on TV

Since Edwin Hubble it's never been the same
Those pictures of other stars pushed me out of the frame
You never even gave me a proper name, like

Alpha Centauri
Epsilon Tauri
Delta Librae
I'll even take HR 2948[**]
Or Kevin?

You've achieved nuclear fusion—oh well done!
Made helium from a little hydrogen—that's very cute, Earth!
Every second I do that to 600 million tons
If I was Marilyn Monroe you'd be ...
 ... the one I can't remember the name of from One Direction[***]

You should have stopped at Copernicus
Then I'd still be the center of your universe
You say I'm just an average ball gas
I think you're talking out of Uranus[****]

1.4 million kilometers[*****]
That's my diameter
Tell me seriously, with those parameters
Have you ever tried to put a hat on there?

Hip hip hip hooray
I'll be a red giant one day[******]
And your Earth will go up in flames
But in the meantime please join my Facebook fanpage[*******]

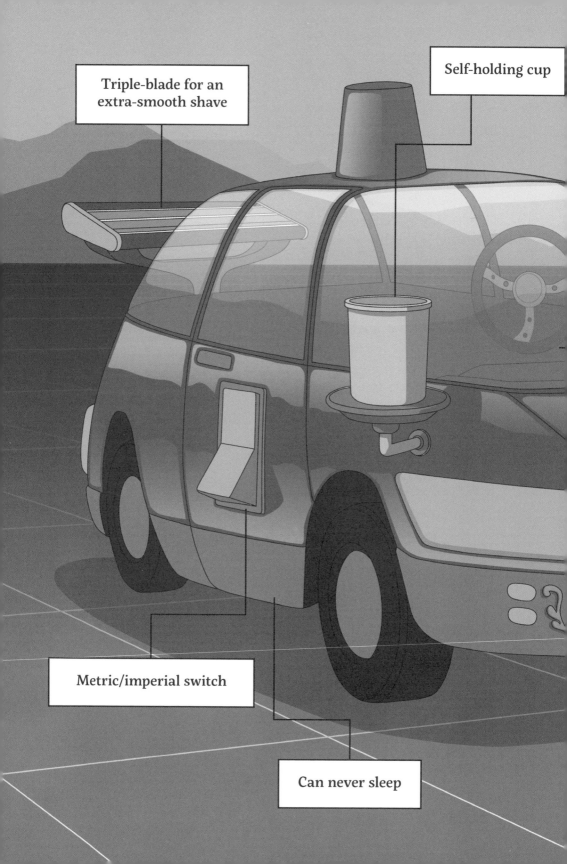

When cut up by another driver, bird is flipped

Decoy steering wheel

Takes 4,096 AA batteries

Plots downfall of puny humans

Original Victorian cornicing

FUTURE STUFF

Here at Festival of the Spoken Nerd HQ, we've gazed into our crystal ball and discovered amazing things like ... the refractive index of quartz. That's very interesting, but it's closer to the predictions of Newton than Nostradamus.

So we've ditched the ball and looked at current and recurring trends instead. From self-driving cars to cryonics, in this chapter we'll see where a few of our favorite bits of technology are taking us, why the mistakes of the past will stay with us into the future and also delve into the misguided predictions of smarter geeks than us.

Can nerds see into the future? Nope, but that's not going to stop us trying ...

SELF-DRIVING CARS

The idea of self-driving cars is not new. In 1972 the government proposed a system for London. It required roads to be modified, or new roads to be built that were essentially like tracks. Amazingly, it actually got built and you can still see parts of it today in east London. It's called the Docklands Light Railway.

In well-controlled environments where you can be sure there won't be any funny business, self-driving vehicles already work very well. For example, self-driving combine harvesters are a staple of modern farming.

BEWARE THE RELENTLESS MARCH OF PROGRESS

Even the surface of Mars is a forgiving environment for automation, as we've shown with rovers. The real challenge is places like Newcastle town center on a Friday night. Newcastle is more confusing than Mars if you're a robot. Or a southerner.[H1]

That's the real revolution everyone is talking about: autonomous vehicles on public roads, artificial intelligence fending for itself in the messy, unpredictable world of the human motorist.

And the reason people are talking about it now is because computers are only just becoming powerful enough to make the complex decisions one needs to be a competent driver.

[H1] Steve is from Newcastle, so he's allowed to say this.

Waymo is currently the biggest player in the driverless car scene, formerly known as the Google self-driving car project. By 2017 their vehicles had clocked up 3 million miles of driving experience. Experience they share with the hive mind, a network of car brains that learn from each other. But it's not just the brains that are important, it's the sensors. And in that regard, Waymo's robo taxi drivers are far superior to humans. As well as high-definition cameras and sensors looking in all possible directions simultaneously, they have spinning lasers that create a detailed 3D model of the world around them and they know the whole of Google Maps off by heart. They even have the equivalent of an inner ear, so they know their orientation in space. Though if you make them do too many doughnuts they will get dizzy.

Waymo's plan is eventually to launch a fully autonomous car brain to the public, but that's not the only route being explored. The other option, considered by BMW and others, is to make and sell cars that are more and more autonomous. For example, they release a car where the cruise control takes care of lane changes. And then one that handles merging too. And so on, getting incrementally better. The problem is, there will come a time when it is almost exclusively driving itself. But occasionally, out of the blue, it will say, "Shit, sorry, I need you to take over. I'm out of my depth." Chances are, at this point, you've switched off and your brain will need to go from 0 to 60 in a few seconds. Scary stuff.

That's Waymo's argument for going all in from the start, and other players are following suit. People do like familiarity, however, and they don't like change. So I propose Waymo also builds a robot that sits in the passenger seat criticizing the decisions of the onboard computer.

What will the future look like? Henry Ford, the inventor of the motor car, once said, "If I had asked people what they wanted they would have said faster horses." Actually, he almost certainly didn't say this, but the point is, we're not very good at knowing what we are going to want in the future. And self-driving cars are a good example of that. We imagine all owning our own self-driving car, but the future probably won't be like that. Instead, you'll subscribe to a fleet of cars and when you need one it will drive up to your house and take you where you want to go. And when you get there you won't have to worry about parking.

The trolley problem

There are arguments against self-driving cars, including moral questions. For example, whose fault is it when a self-driving car injures someone? Or what if the vehicle has to make a choice between continuing straight and killing a pedestrian or swerving and killing the passenger? This is a modern twist on the Trolley Problem, a classic unanswered question in philosophy.

The other argument I hear sometimes is, "But I enjoy driving!" No. You don't. You enjoy the idea of driving. Cars are sold on the dream of the open road but that's not the reality. The reality is sitting in traffic being sworn at.

Some people are concerned about big tech giants running our cars. Google, for example, has some serious rivals, and apparently the first production models will refuse to drive you to the Apple Store.

On the safety front it's a no-brainer: self-driving cars will be safer than the stupid, sleepy, myopic, selfish, angry, and sometimes drunk meat-driven cars of today.

So, yes, there are reasons to be scared of self-driving cars, but once upon a time people were just as scared of self-driving elevators. And in the end the driverless elevator redefined the concept of a city. Driverless cars may well do the same.

YOUR FUTURE IN A FREEZER

What's great about the future is that there's lots of it. And, thanks to the marvels of modern medicine, every generation gets to enjoy even more of that future than their parents. Our life expectancy has doubled in the past 200 years. Some medical researchers predict that the first person to live to 150 has been born already, and projections from the UK's Office for National Statistics show that one in three babies born in 2016 will live to be 100 years old.

This is where it gets personal. Last year I had a daughter, and it's astonishing to think that she has a one in three[H1] chance of becoming a Centurion.

I mean, I had no idea the Roman Legion were still recruiting. Let alone recruiting women![S1]

So this got me thinking ... Is there any way for me to give myself the same life-chances as the next generation?

One option is to pick better parents—and therefore better genes—to increase my potential lifespan. This option means first inventing time travel, and then risking the possibility of not existing at all once I've carried out my *Back To The Future*-style matchmaking in the early 1980s.

Pimping up my current set of genes could be the answer: researchers are looking at ways to disrupt telomeres, a kind of internal self-destruct system that bumps off your cells when they've replicated a certain number of times. Alternatively, identifying and manipulating particular genes to increase longevity seems to be showing positive results in mice, but understanding any side effects—and testing these ideas in humans—is a way off yet.

[H1] Actually, her chances are even higher because, according to ONS's analysis of data from 2014, baby girls have a 35.2 percent chance of reaching 100 and baby boys have a 28.4 percent chance of doing the same.
[S1] It's centenarian, Helen.

Another idea is to exercise frequently, eat well, stay mentally active, participate in my community, and seek out happiness, whatever the situation. Pah! Sounds like effort.

Maybe there's a way to skip all the boring bits and get to the actual future instead, with our bodies still intact?

My husband has a suggestion: we travel at close to the speed of light away from Earth, returning later to find that time has passed faster for terrestrial humans than for ourselves. Nice idea, but to get any sort of meaningful distance into the future, we'd have to spend most of the rest of our lives hurtling through space with only each other for company. No thanks. Not even if the spaceship has Netflix.

Nope, I'd rather get science to offer me a one-shot solution, even if I have to search the fringes of physics to find it. And lo and behold, here it is: I'll get my body frozen, to be revived at some future point when medical science has picked up the slack. Simple!

"If cryonic freezing is the answer, don't trust the person asking the question"

All right, all right. You still need some convincing. Me too. Here are a few of the pros and cons spelt out.

☑ You'll be in good company. As we all know, Walt Disney was cryonically frozen. So even if the future Earth is a barren, postapocalyptic wasteland,[H1] you'll have someone there in the revival pod who can spin a good yarn.

☒ Unfortunately, that's not true: "Disney on ice" is a fiction. Probably the most famous person to be cryonically frozen

[H1] This would be a postapocalyptic wasteland with strangely excellent healthcare facilities.

is baseball legend Ted Williams in 2002. The earliest was psychology professor James Bedford in 1967. Of those that immediately followed suit, some are in a better state than others: a vacuum pump failed at an early storage facility, allowing several of them to thaw out. It's enough to put you off watching *Frozen* ever again.[H1]

☑ At the rate medical science is improving, if you wait long enough you'll not only be cured of whatever you died of, but everything else that's been bothering you in life. That patch of dry skin on your elbows? Good as new. That ankle which twists too easily when climbing stairs? Sorted. That feeling of loneliness when you realize everyone you've ever loved is long gone? Sorry, can't help with that.

☒ Legally, you have to be declared dead before your body can be cryonically frozen. Which is going to mess up your inheritance tax bill at best, and your great-great-great-great-great-great-grandchildren's heads at worst. Obviously, being dead makes the revival process just that teensy bit harder as well.

☑ Committing to cryopreservation might make you feel better. If you don't subscribe to any particular belief system, the hope of revival in the future might be just what you need to keep going in the here-and-now. At the very least, it's a great idea for your next anniversary present. What could be more romantic than a voucher for double cryopreservation?[H2]

☒ It might not work. Remember the frozen strawberries in Chapter 5? When those fruit cells defrosted, all the DNA pretty much fell out with a little help from some pineapple juice and alcohol. It's great for a fruity margarita, not so great for preserving your gray matter. That's why blood is exchanged for a kind of antifreeze before undergoing The Big

ANTI-
FREEZE

[H1] Having spoken to friends with daughters slightly older than mine, I realize this is impossible. I will be watching *Frozen* at least once a day for the next ten years. It's enough to make you long for the peace and quiet of a human-size liquid-nitrogen dewar.

[H2] "Pretty much anything else, ever," says my husband. Looks like I'll have to get him that home-cremation kit for Christmas instead.

PASS THE STRAWS, I'LL HAVE
THAT MARGARITA NOW. LIFE IS
SHORT, ENJOY IT WHILE YOU CAN.

Chill—to limit damage to cells from expanding ice crystals. At a low enough temperature everything then turns solid, which also helps to keep it intact for the future—a process called vitrification. This might work for Canadian wood frogs, who have natural antifreeze in their bodies to survive up to seven months as living iceblocks each winter, but there's no evidence it works in humans on this scale.

But it might work! And if it does, you'll be at the vanguard of the cryonic revolution! The idea that frozen eggs and sperm can become actual real-life people seems pretty normal to us today, although reproductive cells and embryos are a very different proposition compared to fully grown adults. Research into advanced cooling techniques that could help preserve organs for transplant is very much ongoing, and yielding some positive results in the laboratory for small amounts of human heart tissue. And what's a few dollars a day to you? Forego your morning latte and buy an insurance policy instead, to ensure that your body is frozen as soon as feasible after death. Fortune favors the brave, and when those future humans eventually defrost-and-revive, they'll be considered deities among mere mortals. Finally, don't forget that if you sign up for this vanishingly slim chance and it doesn't succeed, at least you'll never find out.

Saying all that, it probably still won't work. And there's no guarantee that your knowledge, memories, or personality will remain intact even if the science eventually works itself out. Not to mention the economics: full body preservation costs upward of $28,000, plus you'll want to set up some kind of long-term investment or trust to fund medical treatments and all the trappings of your postcryonic lifestyle in the future. The more people who peg their hopes on a frozen dream, the less likely it becomes. Imagine half the world's population, and most of the world's currency, tied up in keeping tanks of nitrogen liquid at -320°F ... something's got to give, right?

SH*T WE'RE STUCK WITH— THE FUTURE ISN'T BRIGHT

Walk through the Making the Modern World gallery at the Science Museum in London and you'll see the march of human progress, as one engineering marvel leads on to the next, from the steam engine to the supercomputer. What's missing from the gallery is all the annoying shit we came up with along the way. And quite right too, you might think. Progress weeds out the bad ideas and they're lost to history. Why show the ideas that didn't make it? Except that's not always how it goes down. Sometimes we don't realize an idea is bad until it's too late and we're stuck with it.

Electricity flows backward

Take, for example, the flow of electricity. You might know that an electric current is the flow of tiny particles called electrons,[S1] so when you turn on a flashlight, for example, electrons flow from the battery, through some wire, through the bulb, through some more wire, then back into the other end of the battery, like in this diagram.

[S1] Strictly speaking, electric current is the flow of charge and charge is carried by particles. In the vast majority of cases the particles in question are electrons, but not always. For example, it could be the flow of ions in an electrolyte solution or the flow of "holes" in a semiconductor.

Look at those electrons go! So which way is the electric current flowing? It should be obvious; it should flow in the same direction as the electrons. It should, but it doesn't. Electric current flows in the opposite direction to the flow of electrons. Whose stupid idea was that, you might ask? It was Benjamin Franklin's, but at the time it wasn't stupid, it was just bad luck.

Before Franklin, Charles François de Cisternay du Fay noticed that when glass was rubbed with silk it generated one type of charge and when amber was rubbed with fur it generated another type of charge.[H1] He called the two types vitreous and resinous charges respectively. And, crucially, he noticed that the charges canceled each other out when the amber and glass were brought together.

Franklin proposed a "one-fluid" explanation, suggesting that one type of charge is the result of an excess of fluid and the other type of charge is the result of a deficit of fluid, so that when they came into contact with each other, the fluid would flow from the object with an excess to the object with a deficit. This would explain the canceling out that was observed. Franklin proposed that an excess of fluid should be called positive charge and a deficit should be called negative charge. That way, a flow of positive electric charge—an electric current—would be in the same direction as the flow of the fluid. We now know this "fluid" to be charge-carrying particles. And in the case of rubbed glass and amber, the particles in question are electrons, just like the stuff that flows through electric circuits.

At the time, no experiment had been devised that could determine what the fluid was, or, indeed, in which direction it was flowing. Did the glass have the excess fluid or was it the amber? We don't know why Franklin made the choice he did, but he decided vitreous charge would be positive and resinous charge would be negative; that the glass had the excess fluid and the amber had the deficit. We now know that it's the other way

---

[H1] Yet more thrilling experiments to try at your Static Electricity Party (see page 139).

around. It's amber that holds the "fluid" of electrons. And that's how the electron came to be negatively charged.

Negative charge flowing to the left is equivalent to positive charge flowing to the right, which is why we have the confusing situation that current flow and electron flow in a wire are in opposite directions!

Too many πs in a pie

It's not the only unfortunate idea we're stuck with. Here's an example from the world of mathematics: π! That's right, the mathematical constant π (pi), so beloved by mathematicians, is actually too small.

"Hold on ... what?! π is the ratio of the circumference of a circle to its diameter. Are you saying that *isn't* 3.14159 ...?"

Calm down, you're getting angry. What I'm saying is, π is the circle constant. It's the number we have chosen to represent a fundamental property of circles. But we chose the wrong one! In math, when we talk about (that is, write equations about) circles, we almost exclusively talk about their radius not their diameter. This is partly convention but also because of the way we define a circle—draw a dot on a piece of paper, then pick a distance, like three inches for example, and plot out all the points that are three inches from the dot you just drew.

Congratulations! You've drawn a circle! In fact, you've drawn the *definition* of a circle. A circle is all the points that are some fixed distance from a central point. And that distance is the radius not the diameter. So, the circle constant shouldn't be the ratio of

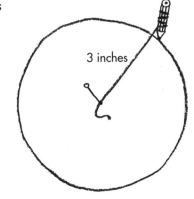

3 inches

the circumference to the diameter, it should be the ratio of the circumference to the radius. Which you might have realized is twice as big as our current value of π. What I'm saying is the circle constant should really be 6.283 …

You might think that this is a lot of fuss over nothing. But it's not.

Making π 3.14159 … makes math harder to do. Like if you measure angles in radians instead of degrees (which you really should, by the way, it's more fun and it's what proper mathematicians do) then a full turn around a circle is a turn through an angle of 2π radians. A turn through half a circle is π radians. A quarter of a circle is $\pi/2$ (or half π) radians and so on.

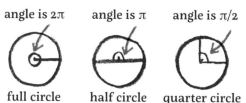

angle is 2π angle is π angle is $\pi/2$

full circle   half circle   quarter circle

The angle is always π times the amount of circle you've turned through *times 2*! This factor of 2 is quite confusing. Seasoned mathematicians have become quite adept at dealing with the inconvenience, but it's a barrier to understanding for people exploring mathematics for the first time. Would it not be better if a full circle was π radians? Then half a circle would be half π radians, a quarter of a circle would be a quarter π radians. How neat! How simple. And that's what you'd get if π was 6.283 …

You'll also notice 2π appearing in equations all the time.

For example, by definition, the equation for the circumference of a circle:

$$C = 2\pi r$$

Or how about the nth complex roots of 1?

$$z = e^{2\pi i / n}$$

It's also in the Fourier transform and the integration over all space in polar coordinates—the list goes on. And I'm not cherry-picking, I promise! How much nicer if all these equations had just π instead of 2π!

What about the formula for the area of a circle?

area = πr^2

If we made π twice as big, the formula would need to be:

area = $\frac{1}{2}\pi r^2$

Not nearly as neat with that ½ in there, you might say. Let's stick with the old definition after all. But actually, the factor of ½ is important as it tells us something about the area of a circle. Consider these other formulas:

Kinetic energy: $\frac{1}{2}mv^2$

Distance an object falls under gravity: $\frac{1}{2}gt^2$

What do they have in common with the area of a circle? They all have a squared term (r^2, v^2, t^2) and they all have that ½ in there. And that's no coincidence. All these formulas can be worked out, or derived, by doing something called integration. When you integrate x, for example, you get $\frac{1}{2}x^2$. So the half tells us something. It tells us that the area of a circle is the result of doing integration. With the standard definition of π, that fact is obscured.

This really is an example of something we're stuck with. There's no way we'd be able to get everyone to agree to change the value of π. We'd have to burn all our old textbooks!

But there is a way out of this mess. All we do is define a new constant, we call it tau (τ) and it's equal to the ratio of the circumference of a circle to its radius, or 2π, and it makes all our equations beautiful and our circle math easier to understand. That way we don't have to replace old books, we just have to slowly introduce our superior constant. This isn't my idea. Check out the great work of Bob Palais and Michael Hartl if you're still not convinced.

RETRO FUTURE-TASTIC!

There are a lot of things that my teenage self thought science would have sorted out by now. Disappointingly, most of them haven't happened yet … or have they?

Take *Jurassic Park*. In the 1993 movie, nature-loving scientists brought extinct dinosaurs back to life, creating a fantasy land of verdant hillsides filled with cute baby velociraptors and the odd flesh-hungry tyrannosaurus rex. If a team of scientists at Harvard have their way, they'll soon be using CRISPR, the latest in gene-editing technology, to tweak elephant DNA and create enough mammoths to populate Pleistocene Park, a Siberian wildlife reserve, where huge four-legged hunks of living history can roam free.

Perhaps, like me, you still have a childish ambition of vacationing on the moon. Just the moon, though? Pah. Get yourself in the queue for Mars One, the organization attempting to land humans on Mars and create a permanent home there. Although applications for one-way "tickets" to the red planet are currently closed, another ambitious commercial spaceflight program—SpaceX—is planning to take the first tourists on a slingshot trip around the moon in 2018. If all goes to plan, it will be the first time humans have traveled beyond low Earth orbit since the final Apollo mission of 1972.

And of course, who could forget the Croydon Atmospheric Railway! A public transport system made out of vacuum-powered tube trains that would have run across my own home turf of southeast London. Planned and tested in the 1840s at a cost of $2.4 million, it never actually opened to the public, mostly because the technology of the time consisted of leather flaps and steam engines. Behold, instead, the Hyperloop! Swapping steam power for levitating superconductors, and Croydon for California, it's basically the same concept back in the saddle. Although there's no guarantee the new version will work, and the price tag is already considerably higher.

There is one piece of fictional technology that still lives on in my heart. It's the one I remember watching on the family's sole household screen, the single cathode-ray-tube television in our living room. A device that seemed impossibly futuristic to my younger self: the *Star Trek* communicator.

The ability to instantaneously speak to someone across the vastness of space, or at least with my best friend at her parents' house ten miles away, was at the top of my teenage wishlist. Yes, it was technically possible to do that back in the 1990s, but relied on both people owning cell phones the size of house bricks, with each device costing approximately the same as one brick house. And those sci-fi series communicators were much more than just chatterboxes, they were portable knowledge banks, geographical locators, and gave you the ability to quiz the ship's computer with your voice. Beam me up!

But how close are we to having them in our hands today?

Actually, pretty close.

Smartphones can do most of it already with a combination of internet access, GPS technology, and voice control. Oh yeah, and they make calls too.

But there's no simple bit of physics that can solve the problem of communicating instantly across the vastness of space. Speed-of-light communication is easy, but it has an in-built limit. On a scale of "standing still" to "infinite speed," the electromagnetic radiation that makes up visible light is still far closer to zero than hero.

Sending a light-speed message to the Starship Enterprise a mere one light year away would still leave you waiting two years for a response. A conversation with a sister-craft orbiting our second-nearest star,[H1] Proxima Centauri, would take more than four years each way. Sending a signal to the moon and back takes 2.5 seconds for the round trip, making conversation pretty awkward. Though it would be even more awkward if you actually found anyone up there to talk to right now ...

[H1] The nearest one being the Sun, of course. Did the guest appearance in Chapter 6 tell you nothing?!?

Some theoretical physicists will tell you that quantum entanglement is the answer to all our problems, but until they can harness the fundamental weirdness of how particles somehow act in a synchronized way from a distance, their ideas are all just theoretical pie in the theoretical sky.

Since *Star Trek* was first aired in 1966, more success has been had with reducing the size and increasing the processing power of computing devices. The world's first portable computer, a teeny-tiny abacus made into a ring, was already making technology wearable 300 years ago in China. Today's cell phones are just another data point in a well-charted curve, which shows that technology is getting smaller and faster all the time: Moore's law.

Not really a law at all, more an observation, it's one that keeps proving itself true year after year. Intel cofounder Gordon Moore looked around the computer hardware industry in 1965 and saw that the number of transistors being squeezed onto every circuit board was doubling each year.[H1] From then until now, the evidence backs up his hypothesis, more or less:[H2] the performance of computer chips has indeed doubled about every 18 months.[H3]

[H1] He adjusted this in 1975 to doubling every two years, but the general idea still holds true.
[H2] I wanted to spell this as "moore or less" but Steve wouldn't let me.
[H3] Although the parallel truism, Wirth's Law, states that software becomes exponentially slower to run as computer power increases, so the net gain from upgrading your hardware is much smaller than you'd expect.

This exponential growth of computer power is a big part of why something closely approximating a sci-fi communicator is in the palm of your hand today. But what's next? Can this trend for smaller, faster, more powerful devices continue forever?

There are some indications that Moore's law may be leveling off, and the end point is obvious when you think about it. Squeezing more and more transistors onto every square inch of circuit board will eventually mean that each transistor gets close to the size of an individual atom. At around five nanometers— that's five billionths of a meter—electrons flowing through a miniature circuit could "tunnel" into somewhere they shouldn't be and cause havoc. Even Intel has admitted that they expect the silicon chips we rely on to reach their limit around 2020.

At that point we'll need a completely new type of circuit board to keep up the pace. Or we'll just have to hope that artificial intelligence evolves better programming skills than humans, in which case the computers themselves can pick up the slack. Perhaps those theoretical physicists and their tangled particles will have solved all the practical problems standing in their way by then, and booted up quantum tunneling technology? Fingers crossed. And uncrossed. Simultaneously.

If you can't wait that long, there is a near-future workaround to getting your own genuine sci-fi communicator. You can now buy a handset that faithfully reproduces the *Star Trek* prop and connects to your cell phone via Bluetooth.

I bet Moore didn't see that one coming.

210

CHOOSE YOUR OWN ENDING (THE KNOWN UNIVERSE EDITION)

You are the Universe.

You are gradually expanding into the infinite darkness.

You are 13.8 billion years old,[H1] give or take 100,000,000 years.

You are bored.

It takes it out of you, all this constant inflation.

"When will it ever end?" you wonder, and sigh, causing a large binary star system to explode. The double supernova sends gravitational waves rippling out across your surface.

"Oops!" you say.

And then do it again for fun.

It looks pretty, after all. And it passes the time.

Turn to the next page.

[H1] You don't look a day past 13 billion, honest.

211

You are now 21 billion years old.

You are still expanding into the infinite darkness.

The humans have gone quiet. In fact they went quiet a while ago, but you didn't really notice at the time. For millennia you'd been blanking out their inane chatter about which bit of rock belongs to whom, why is the frozen white stuff turning into liquid faster than it's supposed to, where has all the light gone, why is there no food anymore, who are we going to eat first, why is the Sun getting so big and red. That sort of thing ...

You never paid much attention when they were there.

Now, suddenly, in the vastness of your own space, you miss them.

You wait to see if they'll start up again somewhere else in your immeasurable volume.

...

...

...

They don't.

You are really, really bored.

Turn to the next page.

212 You find a copy of *The Element In The Room* by Festival of the Spoken Nerd. It's just there, floating along, somewhere near the outer spiral arm of what was once known as the Milky Way. Your curiosity is piqued.

You read about Helen's experiments with homemade litmus noodles.

Excitement sparks in your eyes for the first time in millions of years.

You check the kitchen cabinet.

There is a package of egg noodles in there.

"This is going very well," you think. "Just need some turmeric and we're in business!"

You look further into the cabinet.

There is no turmeric. Science experiments are off the menu, for this eternity at least.

It makes you really sad. A dozen planets tumble out of their orbits, crashing into their sun in a spectacular fiery display.

In Chapter 6 you get to Steve's section about entropy.

You realize for the first time that your inevitable fate is a slow and tedious march toward heat death.

It makes you really, really sad. A constellation of stars collapse together into one supermassive black hole as you contemplate your eventual demise.

You wonder if you can do anything to change your destiny …

Turn to the next page.

213 "Screw this," you think. "I'm the Universe! I can do what I like."

Do you:

Convert all your dark energy to dark matter? *Go to page 213½.*

Cut your speed of light in half? *Go to page 214.*

Increase the mass of your Higgs boson? *Go to page 215.*

Up the strength of your dark energy? *Go to page 216.*

Do nothing? *Go to page 216½.*

213½

You've always been suspicious about that dark energy you've been carrying around all this time.

It's probably what's making you expand the way you do, on this inexorable journey toward heat death.

"Maybe by changing it to dark matter I can slow things down?" you think. "Possibly even reverse this tedious expansion. Anything's better than this!"

So, instead of 68% dark energy, 27% dark matter, and 5% visible matter, you become 95% dark matter.

For a moment, your pulse and your pulsars beat faster in expectation.

You're still expanding. But expanding a bit slower than before.

"Oh great," you say. "I'm still heading for this dumb heat-death ending, and now it's going to take even longer?"

Do you:

Turn that extra dark matter back into dark energy? *Go to page 213.*

Conjure up some more dark matter from somewhere? *Go to page 217.*

Give up and settle in for the long haul? *Go to page 216½.*

214

Nothing changes.

Everything is now traveling at half the speed it used to, or not, depending on your reference frame. Your physical laws and measurements have simply adjusted to accommodate the change. It's basically window dressing your problem, not solving it.

You realize you should have listened to that Einstein chap when he was still around. He tried to get in touch to ask some questions. You never returned his calls.

You're still on the path to heat death, but now it'll take even longer.

"Great," you think, "that's really made my day."

You think that maybe what you need is a more dramatic result. To get it, you must adjust something more essential, more fundamental. Something that isn't defined by a unit of measurement.

Something like the fine structure constant: the strength of the electromagnetic force within you. A dimensionless constant! That should do it.

This electromagnetic force is what holds your atoms and molecules together. Surely changing that will do something? But you're not sure what.

Sounds risky.

Do you:

Chop your fine structure constant in half? *Go to page 217½.*

Just kick yourself for making such a rookie error about the speed of light? *Go to page 213.*

215

You realize—too late—that you exist in a metastable state.

The mass of the Higgs boson is finely tuned to keep you exactly as you are right now.

So finely tuned, in fact, that even a small change in its energy will create a catastrophic event that destroys you.

As the Higgs field breaks down, a bubble containing a new vacuum state comes into being.

It expands at the speed of light, obliterating everything in its path.

As the bubble absorbs you entirely, one of your cosmic eyebrows raises a little and you say, "Well, that was unex …"

You never get to finish your sentence.

You succumb to the Big Slurp.

In a parallel universe, you go back to page 210.

The end.

216

You begin to expand at a greater rate than ever.

All your matter, from the center to the edge, is flung outward, farther and faster than before.

You can't control it. You puff up like a piece of cosmic popcorn.

Galaxies are destroyed. Stars are broken into pieces. Planets explode.

Something is definitely happening.

You are 36 billion years old.

You succumb to the Big Rip.

Go to page 217½.

216½

You continue to expand, ceaselessly.

Galaxies drift apart into an endless night. Your stars twinkle and expire.

All the matter you contain, which had so uniquely clumped together into suns, planets, black holes, and cosmic dust, becomes perfectly evenly distributed across the vastness of you.

Thermodynamic equilibrium is reached.

Entropy is maximized.

You succumb to the inevitable ennui of heat death.

You are ten thousand trillion trillion trillion trillion trillion trillion trillion trillion years old. Approximately.

"This," you think, using your very last clump of energy, "has been incredibly boring."

The end.

217

Your expansion slows.

"Now we're really getting somewhere," you think.

Not content with just slowing down, your expansion stops, and your outer edges begin to shrink back in.

As you collapse in on yourself, galaxies squash into one another, stars smoosh together, planets are crunched up like Tic Tacs as your matter pulls together into a superdense state.

The force of gravity pulls you ever tighter, into a black hole singularity of infinite density.

You succumb to the Big Crunch.

Moments later, you explode in a new Big Bang.

You wait around for 13.8 billion years, give or take 100,000,000 years.

Suddenly, you become self-aware.

Go to page 210.

217½

Physical forces cannot contain your matter.

You are pulled apart in your entirety, blown into smithereens.

Individual atoms are ripped to shreds. Matter of every type is broken down.

You transform into a uniform soup of subatomic particles and nothingness.

"Thank goodness that's over," you say.

The end.

ACKNOWLEDGMENTS

Writing a book is hard. But lots of people made it much, much easier. Thank you to the Octopus team, led by Romilly "two pints of wine" Morgan, without whom this book would not exist. Her boundless enthusiasm, encouragement, and hard work made it all happen. Thank you also to Juliette Norsworthy for creative design and epic page-wrangling skills. Thank you to our editor Pauline Bache and copy editor Charlotte Cole for tidying up our mess into something that could be turned into an actual book. Thank you to illustrators Richard Wilkinson, for his beautiful and funny visual interpretations, and Grace Helmer, who has brought pictorial clarity to our pages. Thank you to Matthew Grindon and Karen Baker for shouting about the book in a structured and effective way, and Kevin Hawkins for getting it onto every conceivable shelf, both physical and digital. Continuing thanks go to everyone at Noel Gay for nurturing Festival of the Spoken Nerd so well. A special thank you to Sophieclaire Armitage at Noel Gay and also Danielle Zigner at LBA, for setting everything up for this book, putting the wheels in motion, dotting the i's and crossing the t's.

There are also a huge number of people who helped us with the words and ideas that have ended up on these pages. Thanks to the midwives at Lewisham Hospital for their kindness and skill, and for putting up with Steve's analysis. And thanks to Ian Lake for making a contractions timing app with an export function. Thanks to Juhani Kaukoranta for making a website that could calculate the position of a constellation so Steve didn't have to. Thanks to Mark Warner and John Biggins for figuring out why the beads go up before they go down. Thanks to Colin Wright, Andrea Sella, and Andrew Pontzen for their endless enthusiasm for sharing interesting stuff. Thanks to Kat Arney for her unwavering encouragement and eagle-eyes. Thanks to Ed Yong for knowing more about gastrointestinal bacteria than is strictly healthy for one person. Thanks to Tom Whyntie for introducing us to the concept of the Banana Equivalent Dose. Thanks to Robin Ince for first treading some of the paths that we follow along, and *Infinite Monkey Cage* producer Sasha Feachem for inviting our special guest "star" on to the radio for the first time. Thanks to Amy Hopwood at Phil McIntyre Entertainments for helping us do what we do all over the UK.

Helen would like to thank her husband Rob and daughter Matilda, who is probably wondering where "Mommy's silver tappity-tap-tap box" has gone, now this book is finished.

Steve would like to thank his family: Lianne, Ella, Lyra, and $newbabyname for being so supportive during an often grueling creative process.

And finally, thanks to our friend and colleague Matt Parker for being the mathematical rigor behind Festival of the Spoken Nerd and for his long-standing support of the Tau Movement. We didn't have room to print your appendix. Sorry, not sorry.